Cover: F. Schmidt/Fotolia

Bibliographische Information der Deutschen Bibliothek
Die Deutsche Bibliothek verzeichnet diese Publikation in der Deutschen Nationalbibliographie; detaillierte bibliographische Daten sind im Internet über http://dnb.ddb.de abrufbar.

Michael Dornbusch, Ulrich Christ, Rob Rasing
Epoxy Resins: Fundamentals and Applications
Hanover: Vincentz Network, 2016
European Coatings Library
ISBN 3-86630-887-6
ISBN 978-3-86630-887-9

Please ask for our book catalogue
Vincentz Network, Plathnerstr. 4c, 30175 Hanover, Germany
T +49 511 9910-033, F +49 511 9910-029
books@european-coatings.com, www.european-coatings.com
Satz: Vincentz Network, Hanover, Germany
Druck: BWH GmbH, Hanover, Germany

ISBN 3-86630-887-6
ISBN 978-3-86630-887-9

European Coatings Library

Michael Dornbusch

Ulrich Christ

Rob Rasing

Epoxy Resins

Fundamentals and Applications

Foreword

Hardly any class of resins is more widely used in the coatings industry than that of epoxides. Therefore, anyone who works with coatings needs a good overview of this topic.

This book seeks to provide that very knowledge from three aspects:

First, it surveys the historical development of epoxy resins, from the first synthesis in the 19th century to the first patents for epoxide compounds in coatings to industrial scale production of resins based on bisphenol A. Chapter 1 also proposes a unique nomenclature based on DIN standards for avoiding the proliferation of terms used for epoxy resins in the coatings industry.

Second, the book contains a compilation of the chemical properties of the epoxy (oxirane) group, of polymers containing epoxy groups and of polymers produced with epoxy groups (phenoxy resins). Chapter 2 presents the chemistry of the epoxy group, i.e. a characterisation of the three-membered ring, followed by the chemistry that is facilitated by this functional group and that is actually used within the industry. Finally, it examines the production and characterisation of polymers, both with and without epoxy groups, along with their curing reactions.

A comprehensive overview of all the possible reactions of epoxy groups is provided, with the focus on reactions that are relevant to coatings. Modern methods of characterising the compounds, such as NMR, IR and NIR spectroscopy are explained on one hand, while all the typical key parameters used in industry and based on corresponding DIN and ISO standards are presented on the other.

The third aspect is coating agents and how they are used in industry.

Chapter 3 covers the general use of epoxy resins and phenoxy resins. Sections 3.2 to 3.5 discuss the state of the art regarding the use of epoxy resins in industrial areas such as corrosion protection, flooring, powder coatings, can and coil coatings, as well as offering sample coating formulations and discussing the property profiles of the resulting coating surfaces. The topic of can coatings, in particular, includes a detailed discussion of the toxicological properties and current legislation on the use of resins based on bisphenol A.

Chapter 4 deals with trends in the epoxy resins market. Among the topics here are new applications, alternative compounds for BPA in food packaging, and new trends in the coatings industry.

This book will serve not only as a reference book on the chemistry of epoxides and their properties, but also as a monograph on the industrial coatings applications of epoxy resins, both with and without epoxy groups. It will therefore prove useful to students, developers and industrial users alike.

Düsseldorf, December 2015

Michael Dornbusch

Contents

1 Introduction

Michael Dornbusch

1.1 History

The history of epoxy resins began in 1854 when *Berthelot* first prepared epichlorohydrin by making glycerol react with phosphorus trichloride [19–21].

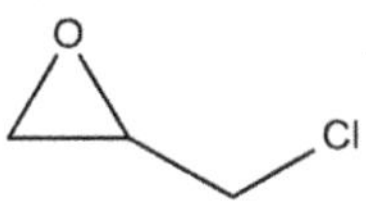

Equation 1.1: Chloromethyl oxirane (epichlorohydrin)

The next step occurred in 1891 with the first description of 2,2-bis (4-hydroxyphenol)-propane (bisphenol A) by *Dianin* [2], who produced the impure compound [10].

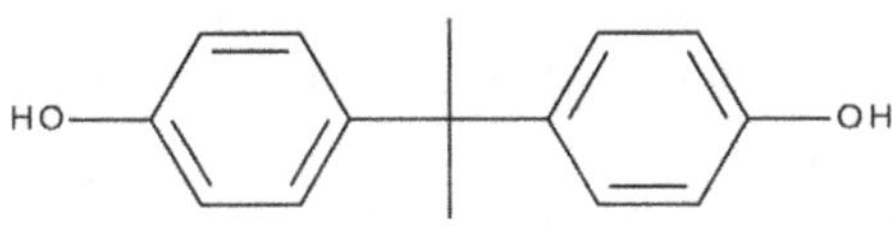

Equation 1.2: 2,2-Bis-(4-hydroxyphenyl)-propane (bisphenol A)

Sixteen years later, in 1905, Zincke in Marburg, Germany, synthesised pure bisphenol A (BPA) from acetone and phenol [10].

In 1909, the Russian chemist *Prilezhaev* converted numerous olefins into epoxides by reaction with peroxybenzoic acid [8, 22].

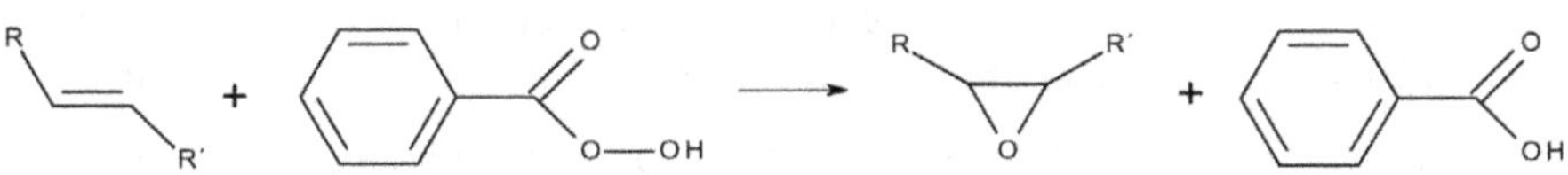

Equation 1.3: Preparation of epoxides, according to Prilezhaev [22]

A patent published in the same year by *Horn* [9] claimed protection for a protein-based coating formulation, which was obtained by mixing epichlorohydrin and proteins, e.g. protalbin or albumose, in a ratio of 1 : 1 in an alcoholic solution. When linseed oil was added to this solution, no turbidity occurred. This was probably the first patent for an epoxy-based coating formulation.

The most-commonly cited inventor of epoxy resins is *Schlack* [15], who claimed protection in his patent for I.G. Farben in 1934 for the reaction between bisphenol A and epichlorohydrin to yield epoxy resins [3, 8]. The resins were cured with ethylene diamine. Thereafter, the coatings industry intensified its development activities in the field of epoxy resins.

Patents obtained by *Castan* in 1938 for the company De Trey AG, Switzerland, described the production of a resin, which was based on BPA and epichlorohydrin and was cured with phthalic anhydride. This curing process was done stepwise to yield pre-cured casting resins, which cured after application [5].

De Trey AG produced epoxy-based resins for dental applications [3] but was unable to bring the products to market [18].

Also in 1938, *Stein and Flemming* from I.G. Farben patented an improved synthesis for epichlorohydrin [17] that facilitated the commercialisation of epoxy resins by substantially boosting the yield.

In 1939, *Bock and Tischbein* from I.G. Farben [14] patented the reaction between diepoxides and polyamides and used the resulting compounds for textile applications.

A patent by *Castan* in 1943 described the use of catalytic quantities of bases to effect curing [6]. In the USA in the same year, *Greenlee* [16] patented the resin obtained from the reaction of BPA with epichlorohydrin and its use for coating applications, thereby laying the foundations of the industrial use of epoxy resins in that country.

Industrial production of bisphenol A from acetone and phenol started after 1945 (1946 according to [18]) [2]. In Europe, Ciba AG developed products under patent licence from De Trey AG that it sold under the trade name Araldite while, independently in the USA, the Devoe & Raynolds Company developed similar products [3].

Commercialisation of the resins by Ciba AG in Europe and by the US companies mentioned above led to a continuous rise in epoxy resin production after 1947 [8].

In the late 1940s, Shell and Bakelite Co. (later: Union Carbide Corp.) commenced R&D activities in the field of BPA-based epoxy resins [18]. At that time, Shell was the sole producer of epichlorohydrin and Bakelite was one of the largest producers of phenolic resins and BPA [18].

In the 1950s, BPA was also used to produce polycarbonate [2], and this increased global production of BPA.

In 1955, a cross-licensing agreement among the four US producers of epoxy resins saw Dow Chemical Co. and Reichold Chemicals Inc. enter the market when they joined the patent pool [18].

In the 1960s, the range of epoxy resins on the market surged dramatically. Ciba AG produced epoxidised o-cresol-novolak resins, Dow Chemical Co. offered epoxidised phenol-novolak resins, Shell introduced multiply epoxidised tetra-functional phenols and Union Carbide entered the market with multiply functionalised epoxides in the form of triglycidised p-aminophenols [18].

Also in the 1960s, Ciba AG in Europe and Union Carbide in the USA established industrial production methods for the epoxidation of olefins with peracetic acid by the Prilezhaev reaction. Ciba AG launched cycloaliphatic epoxy resins onto the market in 1963, following these up with additional products in 1965 based on licences obtained from Union Carbide [18].

Finally, in the 1970s, Ciba-Geigy AG developed epoxy resins based on hydantoin and Shell developed resins based on hydrated bisphenol A, but both product groups had little success on the market [18].

The hormonal activity of BPA, now considered a toxicological property, has a historical background, too [4]. The British chemists *Dodds* and *Lawson* [11, 12] were searching in 1936 for chemicals that would make suitable replacements for natural oestrogen in medical treatments.

A bio-assay revealed that bisphenol A was a substance with a weak oestrogenic effect. It was then discovered that derivatives such as diethylstilbestrol [13] were much more potent and so BPA never found use as a drug [7]. These results have since been confirmed several times [4], but there is controversy surrounding the implications.

1.2 Applications for epoxy resins

Outstanding properties, such as resistance to humidity and chemicals, good adhesion to numerous substrates, and good mechanical properties combine to make epoxy resins versatile construction materials and coating agents [18].

Applications for epoxy resins can be divided into three areas:

- Coatings
- Adhesives
- Construction materials.

A rough overview of these is given below.

1.2.1 Coatings

The best known application in this area is likely to be that of heavy duty corrosion protection (see Section 3.2) [8]. Major examples here include shipbuilding, offshore, and engineering structures, such as bridges, with solvent-borne, water-borne and solvent-free coatings, cured with various amines, being used in all areas [8].

The automotive industry uses resins based on epoxy-amine adducts that have been produced from epoxy resins (see Section 3.1.3). Resins bearing amine groups can be protonated and the resulting cationic particles deposited by means of cathodic polarisation on a workpiece [8] and cured with blocked isocyanates at 170 to 190 °C, i.e. they can be covalently crosslinked. This cathodic electrodeposition coating (or E-coating) process provides the corrosion protection found on modern car bodies (see Section 3.1.3).

Another important application area is that of powder coatings (see Section 3.4). When epoxy resins are combined with suitable hardeners, such as dicyandiamide (DICY) (see Section 2.1.1), acid anhydrides (see Section 2.4.2), phenol novolaks (see Section 2.4.1) or polyisocyanates (see Section 2.1.1), the outcome is thermosetting powder coating systems that possess outstanding properties [8].

One of the oldest applications is that of internal can coating [8]. Owing to their strong yellowing, these epoxy-resin-based coatings are also called "gold coatings" (see Section 3.5) and are cured with cresol resols at elevated temperatures [8].

Epoxy resins are also successfully employed in specialty applications, such as UV-curable epoxy resin systems (see Section 3.1.2) in UV-curable solder resists and protective coatings for printed circuits, especially for fine-line and multilayer boards [1].

This list could be continued indefinitely, not only as regards applications for epoxy resins but also combinations with other resin types. Foremost among these are alkyds etherified with epoxies, polyacrylic resins that react with the OH groups of the epoxy resins, and amino resins, such as melamine, which are able to react with epoxy resins in different ways [8].

1.2.2 Construction materials

Applications in construction materials can also be divided into two main groups, the first of which combines epoxy resins with other materials (fibres) to make construction components. The other uses epoxy resins in electrical and electronic engineering.

The main application area for epoxy resins as matrix materials is that of composites. "Composites are always the best choice when a combination of properties is needed that one material cannot provide on its own" [1]. Fibre-reinforced epoxy resins are composites which are combined, e.g. with glass-fibre reinforcements, to produce aircraft parts and blades for wind turbines [1], i.e. epoxy resin composites have established themselves particularly in lightweight engineering applications.

Epoxy resins are also combined with other materials like graphite, boron or Kevlar fibres [18] to generate materials that have high-precision property profiles.

Wide-ranging applications for epoxy resins are to be found in electrical and electronic engineering.

In electronics, they serve as conformal coatings or laminating resins for the base material of printed circuit boards [1].

The printed circuit boards found in almost every electrical device consist of fibre-reinforced epoxy resins coated with copper. The epoxy resins are cured with dicyandiamide (DICY), amines or imidazoles [18].

In general, applications in electrical and electronic engineering are dominated by curing with anhydrides [1] because this kind of application benefits particularly from the low viscosity, long pot-life and low exotherm [1].

Electrical engineering has been using epoxy resins for 60 years, i.e. just after industry found applications for them. Most uses are in transformers and insulators [1]. An excellent overview of this topic is provided by *Möckel* and *Fuhrmann* in their book "Epoxidharze" [1].

1.2.3 Adhesives

A strong bond between two identical or different materials, such as metals, glass, ceramics, wood, fibres and many plastics, can be obtained with adhesives based on epoxy resins [18]. The various applications, raw materials and processes are presented in detail in Section II-2 of "Formulating Adhesives and Sealants" by *Müller* and *Rath*.

1.3 Terms and markets

1.3.1 Nomenclature

The nomenclature of epoxy resins is confusing, because different designations and colloquial terms are used in parallel.

Epoxides contain epoxy groups, i.e. three-membered rings with an ether function (see Section 2.1). The International Union of Pure and Applied Chemistry (IUPAC) and Chemical Abstracts (CA) call these oxiranes [23]. However, this systematic designation has not become widely established, especially in the coatings industry, which continues to favour the terms epoxy resins and epoxides.

Epoxides found in industry are mainly produced from epichlorohydrin, giving rise to a methyloxirane group, known as the glycidyl group. Glycidyl ethers or esters are the most commonly employed compounds thereof [23].

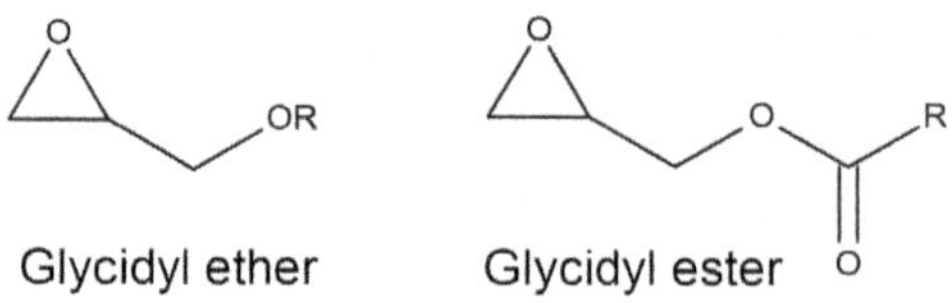

Equation 1.4: Typical oxirane compounds found in epoxy resins

ISO 7142 defines an epoxy resin as a "synthetic resin containing epoxy groups generally prepared from epichlorhydrin and a bisphenol" while DIN 16945, with regard to reactivity, states that "epoxy resins are reaction resins containing sufficient epoxide groups for curing".

The most accurate definition is given in [8], which makes reference to DIN 7728: "epoxy resins are oligomeric compounds containing more than one epoxide group per molecule". These examples alone give some indication of the variation in definitions and nomenclatures employed.

The classification given in ISO 3673-1, which places epoxy resins into classes, is covered in Section 3.1. Finally, some resins that do not contain any epoxy groups are also called epoxides. These resins are polyether polyols, which are mainly synthesised from epichlorohydrin and BPA and which have no detectable epoxy groups in the molecule but are produced from epoxy groups [8]. This class of resins is known as phenoxy resins (see Section 2.4.1) and will be referred to as such in this book.

1.3.2 Markets

The growth of coatings production in Germany in the last five years is shown in Figure 1.1.

The impact of the 2008/2009 global economic crisis on coatings production volume is clearly visible, but so also is the fast recovery in the following years.

In contrast, the crisis had no visible effect on production volumes of epoxy resins for the adhesives market and for coatings (Figure 1.2). Only production of waterborne coatings (Figure 1.3) declined after 2007.

The epoxy resins market may therefore be considered stable. This stands in contrast to the German coatings market as a whole, which has declined since the turn of the millennium. Expressed differently, the coatings market needs economic growth of 2 % in order for it to grow, because the coatings market has been rising more slowly than gross domestic product (GDP) since 2000 [28].

This is a German phenomenon, because the global coatings market is growing at the same rate as global GDP and so is a growth market.

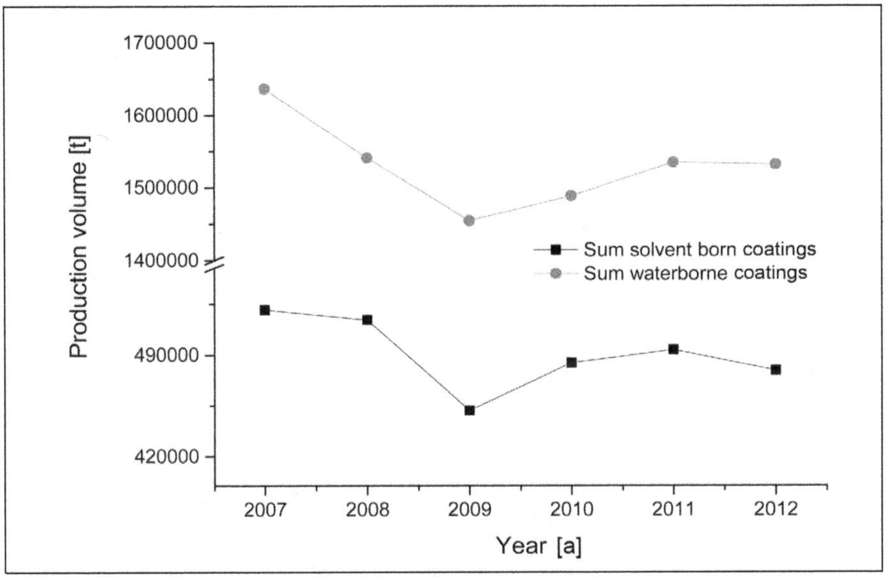

Figure 1.1: Production volumes for waterborne and solvent-borne coatings in Germany [25–27]

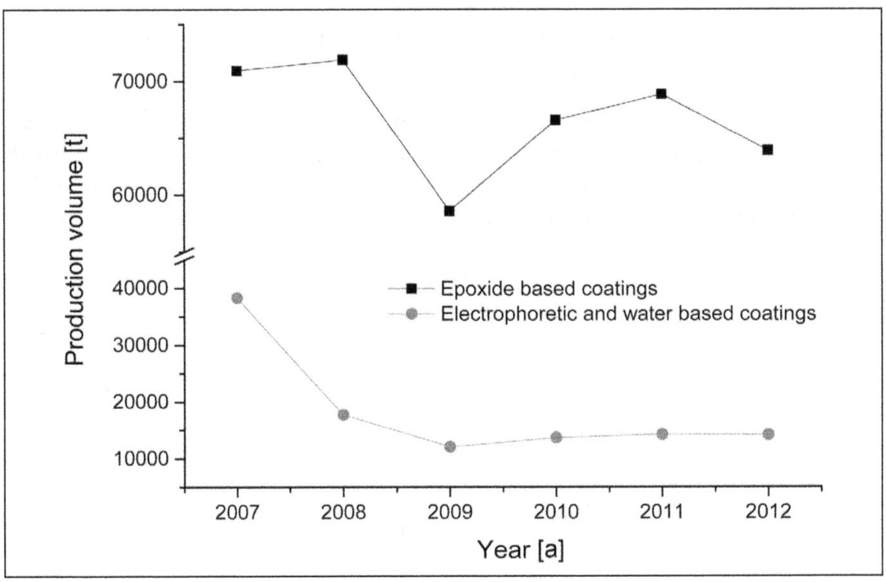

Figure 1.2: Production volumes for epoxy resins in Germany [25–27]*. No data are available for the year 2010.*

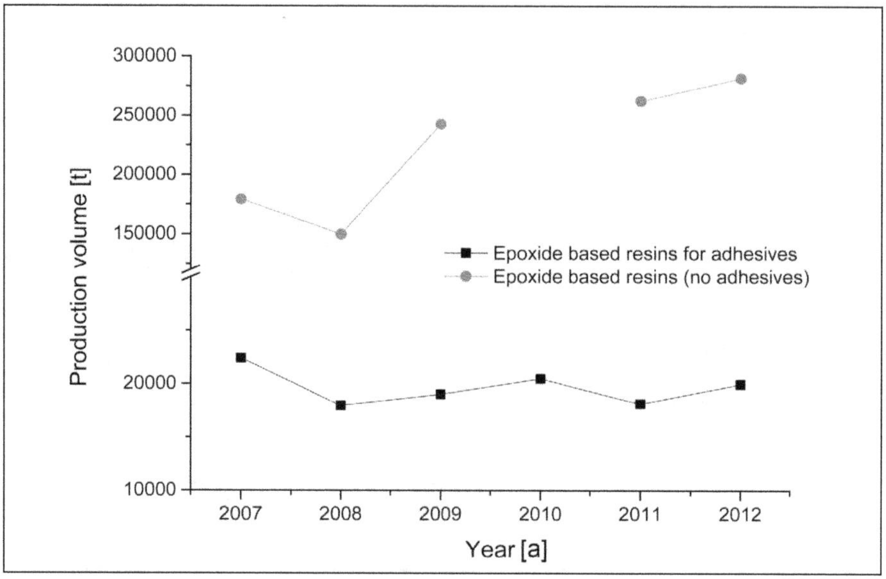

Figure 1.3: Production volumes for epoxy-based coatings and electrophoretic and waterborne coatings in Germany[25–27]

The global epoxy resins market was forecast to increase to 1.93 million metric tons by 2015, according to a 2010 study by Global Industry Analysts Inc. (European Coatings journal, 10-2010).

This growth is being driven by such market segments as "electrical laminates" and "decorative powder coatings". The Asia-Pacific region is the largest growth market in the world, although some plants in this region were shut down during the 2008/2009 global economic crisis.

1.4 Literature

[1] J. Möckel, U. Fuhrmann, Epoxidharze, Die Bibliothek der Technik Volume 51, Verlag Moderne Industrie AG & Co., Landsberg/Lech, 1990
[2] Ullmanns Encyklopädie der technischen Chemie 4th Edition, Volume 18, Petrosulphonate bis Plutonium, Verlag Chemie, Weinheim, 1979, p. 215
[3] Ullmanns Encyklopädie der technischen Chemie 4th Edition, Volume 10, Dentalchemie bis Erdölverarbeitung, Verlag Chemie, Weinheim, 1975, p. 572–580
[4] Thesis, M. Gehring, Verhalten der endokrin wirksamen Substanz Bisphenol A bei der kommunalen Abwasserentsorgung, Technische Universität Dresden, 2004
[5] CH000000211116A und DE000000749512A
[6] DE000000943195B
[7] Umweltbundesamt, Bisphenol A, Massenchemikalie mit unerwünschten Nebenwirkungen, July 2010
[8] H. Kittel, Lehrbuch der Lacke und Beschichtungen, Volume 2, S. Hirzel Verlag, Stuttgart, 1998, p. 267–318
[9] DE000000217508A
[10] Th. Zincke, Justus Liebigs Annalen der Chemie 1905, 343, 75–99
[11] E.C. Dodds, W. Lawson (1935). Molecular Structure in Relation to Oestrogenic Activity. Compounds without a Phenanthrene Nucleus. Proceedings of the Royal Society 125 (839): 222–232
[12] E.C. Dodds, W. Lawson (1936). Synthetic Oestrogenic Agents without the Phenanthrene Nucleus. Nature 137: 996
[13] E.C. Dodds, L. Goldberg, W. Lawson (1938). Oestrogenic Activity of certain Synthetic Compounds. Nature 141 (3562): 247-8
[14] DE000000731030A
[15] DE000000676117A
[16] US000002456408A
[17] DE000000735477A
[18] Encyclopedia of Polymer Science and Engineering, Vol. 6: Emulsion Polymerisation to Fibres, Manufacture, John Wiley & Sons, New York, 1986, p. 322–382
[19] Beilstein Handbuch der Organischen Chemie 4th Edition,Volume 17, Heterocyclische Reihe, Springer, 1938, p. 6
[20] Berthelot, Annales de Chimie et de Physique, 1854, XLI, 299f
[21] Zincke, Justus Liebigs Annalen der Chemie 1857, 101, 67–99
[22] N. Prilezhaev, Ber. Dtsch. Chem. Ges. 42, 1909, 4811–4815
[23] Z.W. Wicks, F.N. Jones, S.P. Pappas, Organic Coatings: Science and Technology, Vol. 1, John Wiley & Sons, New York, 1992, p. 162f
[24] Ullmanns Encyclopedia of Industrial Chemistry, 5th Edition, Volume A9, Dithiocarbamic Acid to Ethanol, VCH, Weinheim, 1987, p. 531–563
[25] Welt der Farben, July/August 2010, p. 38f., Quelle Statistisches Bundesamt, Verband der deutschen Lack- und Druckfarbenindustrie
[26] Welt der Farben, June 2012, p. 28f., Quelle Statistisches Bundesamt, Verband der deutschen Lack- und Druckfarbenindustrie
[27] Welt der Farben, June 2013, p. 26f., Quelle Statistisches Bundesamt, Verband der deutschen Lack- und Druckfarbenindustrie
[28] Farbe und Lack, 7/2013, 119, p. 6–9

2 Basic chemistry of the epoxy group

Michael Dornbusch

2.1 Properties and reactions of epoxy groups

Epoxides (oxiranes) are cyclic ethers that are characterised by high ring strain, which amounts to 114 kJ/mol in oxirane [1, 2] and 106 kJ/mol in oxetane [1]. The values for cyclic hydrocarbons are in the same range: 115 kJ/mol for cyclopropane [1] and 111 kJ/mol for cyclobutane [1].

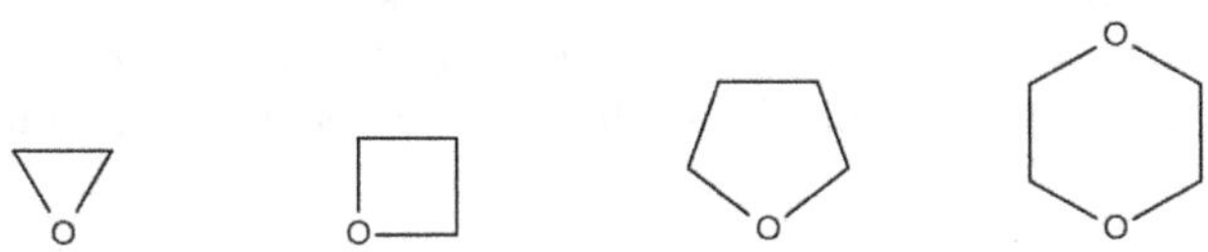

Oxirane (Ethylene oxide) | Oxetane | Oxolane (Tetrahydrofuran) | Dioxane

Equation 2.1: Important cyclic ethers and their systematic and trivial names

The ring strain results from the bond angle of 60°, which is considerably less than the normal tetrahedral carbon angle of 109.5° and the C-O-C bivalent bond angle of 110° in ethers [7].

Small rings are stabilised by attached alkyl groups; thus the ring strain in 2-methyloxirane is 4 kJ/mol lower. The ring strain makes epoxides much more reactive than other cyclic ethers.

The ring strain in oxetane also enables it to react in mild conditions; its reactivity ranks between that of oxirane and open-chain ethers [1]. In contrast, the higher homologues of the cyclic ethers are good solvents and are largely inert.

The key reactions of epoxides can be divided into two groups:

- Reactions with nucleophiles in neutral solution and
- Base-catalysed and acid-catalysed reactions

M. Dornbusch, U. Christ, R. Rasing: Epoxy Resins

2.1.1 Reactions with nucleophiles

As a general rule, ethers are inert to bases, which is why they serve as solvents in numerous organic reactions. By contrast, epoxides undergo ring opening in mild conditions when attacked by nucleophiles, such as alkyl amines, in what is formally an addition reaction without elimination (see Equation 2.2).

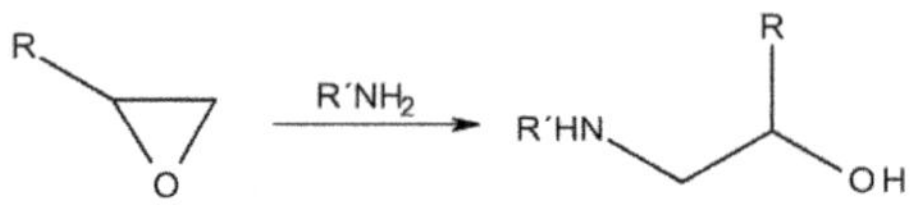

Equation 2.2: Ring opening of an oxirane by a nucleophile

Orientation of epoxide-ring opening

In basic or neutral conditions, the nucleophile attacks the less substituted carbon atom, with ring opening and inversion in an S_N2 reaction [1]; this creates a transition state which polarises the C-O bond, but the point of nucleophilic attack is determined by steric factors [5].

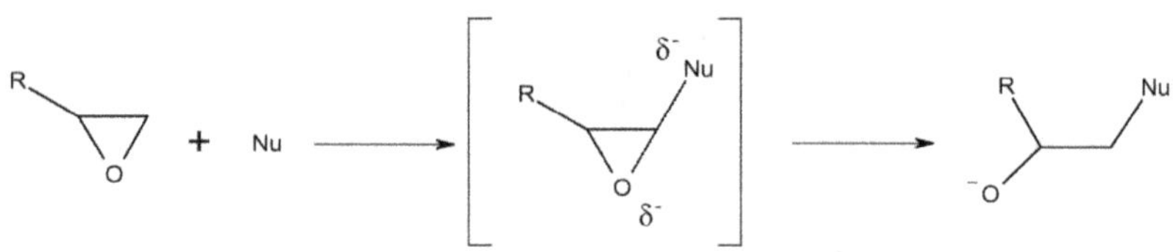

Equation 2.3: Mechanism behind opening of the epoxide ring in a nucleophilic attack

Nucleophilic attack occurs at the less substituted carbon and only one product is generated [2, 3]. A list of important epoxide reactions in basic and neutral conditions that are of relevance to coatings technology is presented below.

Reactions with ammonia, primary, secondary and tertiary amines

Ammonia reacts with epoxides to yield mono-, di- or tri-alkanolamines, depending on the molar ratio of the reactants [3].

The reaction of epoxides with primary amines produces in the first instance a hydroxylamine containing a secondary amine group and an alcohol group. The secondary amine then reacts with an epoxy group to produce a tertiary amine and two secondary alcohol groups [4].

Secondary amines react in a similar way to yield tertiary amines containing a secondary hydroxyl group.

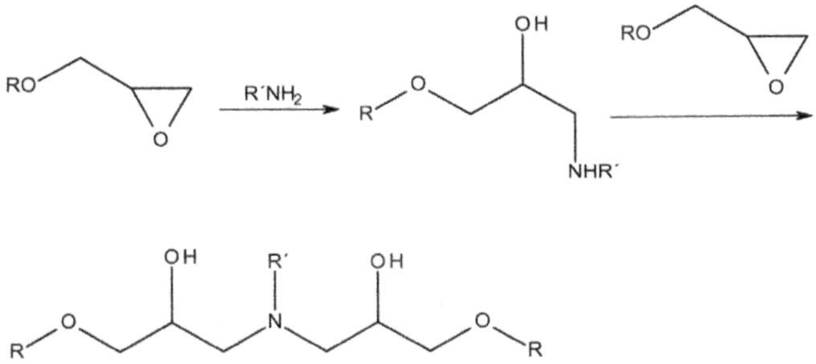

Equation 2.4: Reaction between amines and epoxides

The alcohol group will not react with epoxide to form ether groups if a stoichiometric ratio or an excess of amine is chosen and no catalyst is added [4]. This long-disputed fact was eventually confirmed by intensive spectroscopic analysis of the reaction products and their derivatives obtained at temperatures of 20 to 120 °C [21, 55, 56].

Primary amines react twice as fast as secondary amines with epoxides [4]. Hydroxyl groups can catalyse the reaction between epoxide and amine. Equation 2.5 shows the mechanism proposed for this catalysis [57].

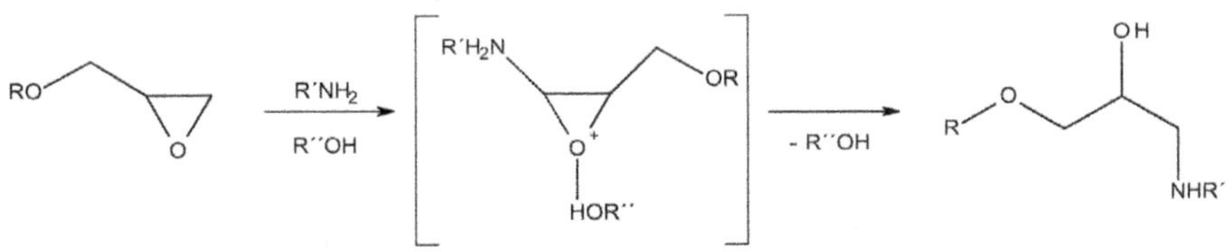

Equation 2.5: Mechanism behind hydroxyl-group catalysis in the reaction between amines and epoxides [4]

Mono-functional aliphatic alcohols exert hardly any catalytic effect, a fact which correlates with the number of OH groups. Poly-functional alcohols therefore exert the greatest effect [4], although phenols also make good catalysts [13] (see Section 2.4.2).

Tertiary amines and tertiary phosphines can catalytically cleave epoxides and both make excellent catalysts for homopolymerisation reactions (see Section 2.4.2 for more details) [3].

At elevated temperatures, tertiary amines react with epoxides to yield quaternary ammonium salts, as shown in Equation 2.6 [13].

Quaternary ammonium salts can also be generated in a *Menshutkin* reaction by making a tertiary amine react with an alkyl halide [58].

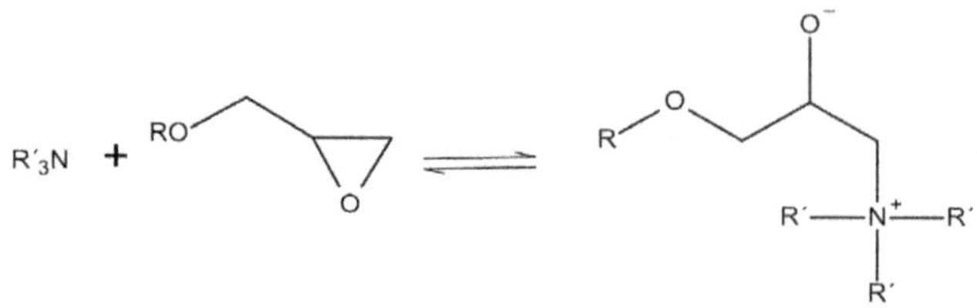

Equation 2.6: Formation of quaternary ammonium salts from tertiary amines and epoxides

Dicyandiamide

In 1862, *Haag* was the first to synthesise dicyandiamide (DICY), a highly polar compound which has a melting point of 205 °C and is insoluble in the low-polarity epoxy resins under ambient conditions. This underlies the use of DICY as a latent hardener [4, 12, 13]. In solution, two tautomers exist [8], as shown in Equation 2.7.

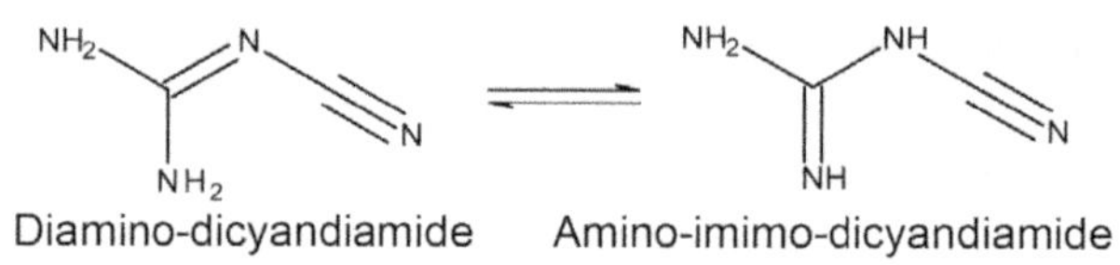

Equation 2.7: Tautomers of dicyandiamide

The concentration of the less polar but more reactive amino-imino-dicyandiamide is so low on account of its sparing solubility in epoxy resins (BPA-based) under ambient conditions that DICY-epoxy resin systems can be stored at room temperature [8].

At 150 °C (model temperature in [8]), a substantial amount of DICY dissolves in the epoxy resin and the concentration of the more reactive tautomer increases, but undissolved agglomerates of DICY are still present in the epoxy resin. Both tautomers react with the epoxy groups.

At 170 °C, the tautomeric equilibrium shifts to the more reactive amino-imino-dicyandiamide, but both tautomeric species still exist and both react with the epoxy groups.

At elevated temperatures, besides the ring-closing reaction postulated by *Gilbert* [11] and *Zahir* [10], open-chain reactions of both tautomers are possible. Zahir's proposed urea formation based on the reaction of the nitrile groups with hydroxyl groups (Equation 2.10) occurs above 170 °C and is not observed at lower temperatures. The formation of urethane groups is also discussed in the literature, but

Gaukler, at least, has been unable to verify this [8]. Likewise, the reactions proposed by *Saunders* [9] and others have not been substantiated by means of infrared spectroscopy [8].

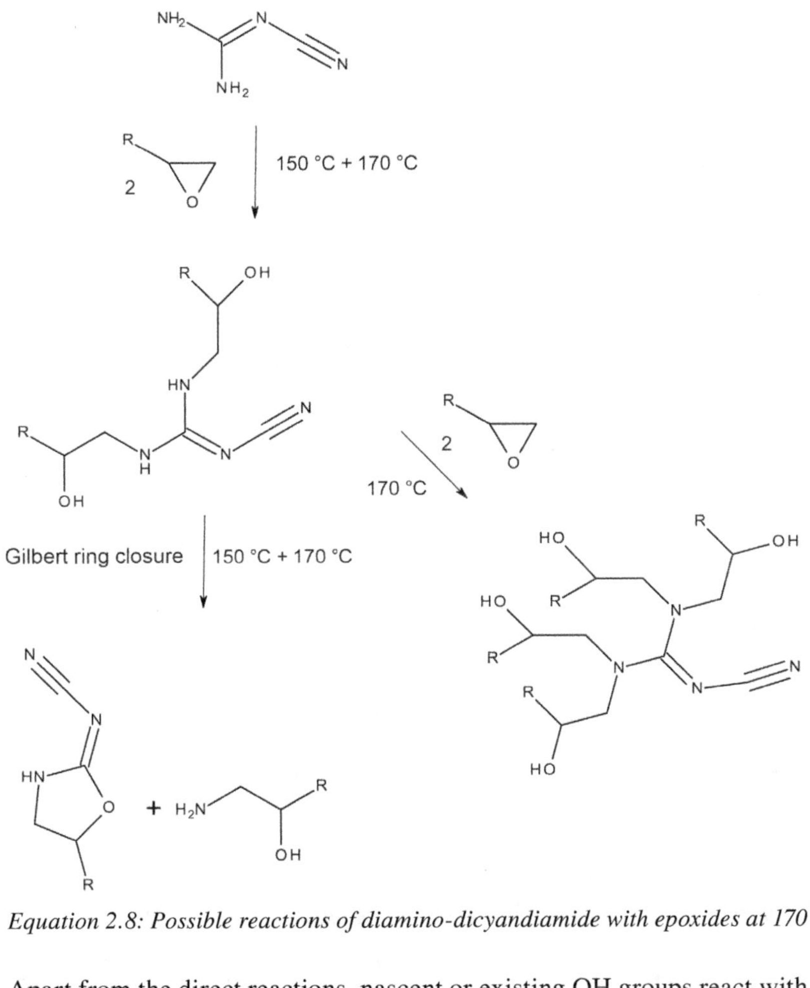

Equation 2.8: Possible reactions of diamino-dicyandiamide with epoxides at 170 °C [8]

Apart from the direct reactions, nascent or existing OH groups react with the nitrile groups at 170 °C (Equation 2.10).

DICY is therefore bivalent relative to the epoxy groups at low temperatures, tetravalent relative to the epoxide groups at elevated temperatures and monovalent relative to the OH groups due to the reaction with the nitrile group. DICY is generally considered to be trivalent relative to epoxy groups [14] because not all of the DICY is dissolved and so some is unavailable to react with them.

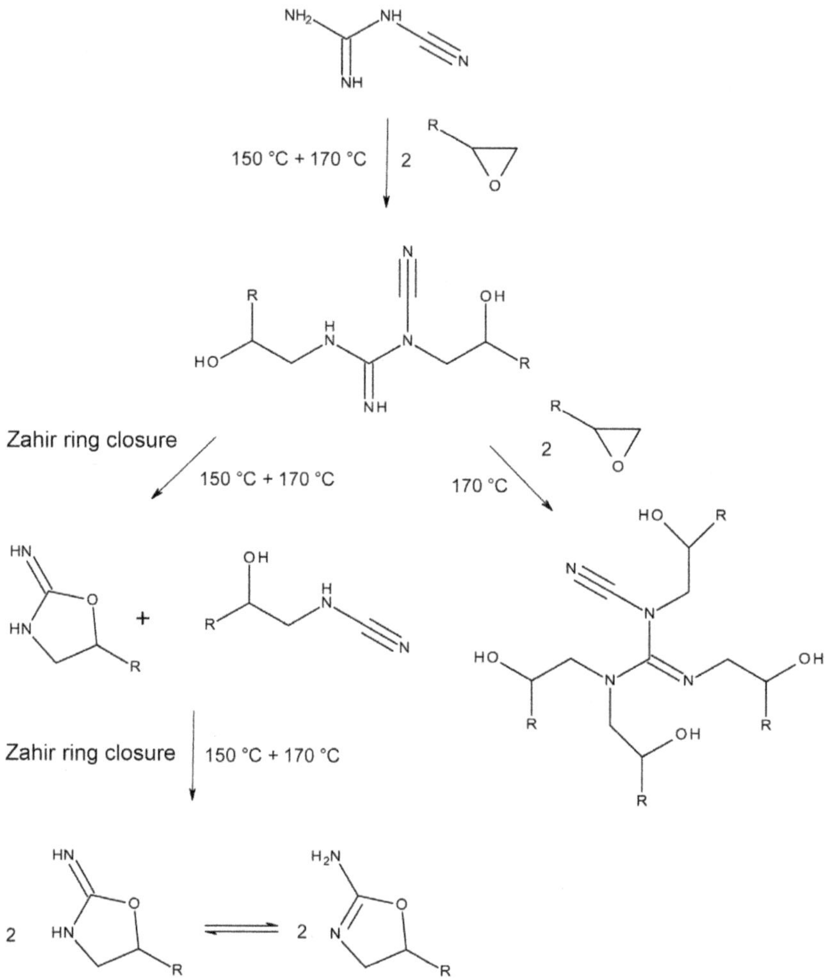

Equation 2.9: Possible reactions of amino-imino-dicyandiamide with epoxides at 170 °C [8]

If accelerators are added to the resin, e.g. 1-methylimidazole, the already complex reaction becomes even more so [8].

Reaction with water, alcohol and phenol

Epoxide groups undergo base-catalysed reactions both with water to yield 1,2-diols and with alcohols and phenols to yield ß-hydroxy ethers of the corresponding regiochemistry [3, 7, 13], as shown in Equation 2.11.

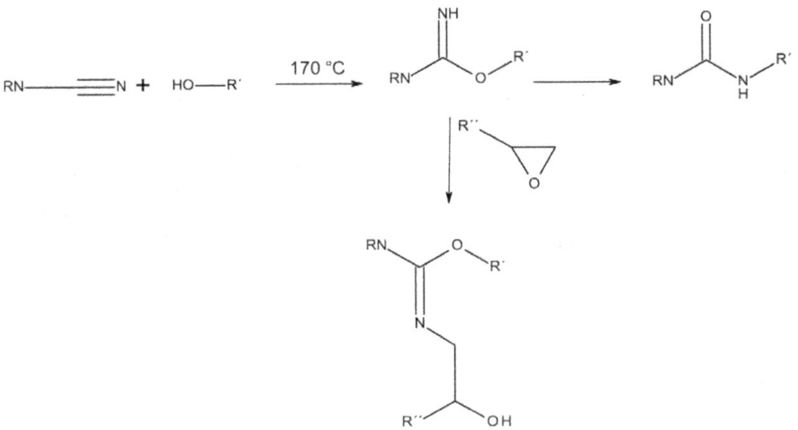

Equation 2.10: Possible reactions of the nitrile group with hydroxides and epoxides [8, 18]

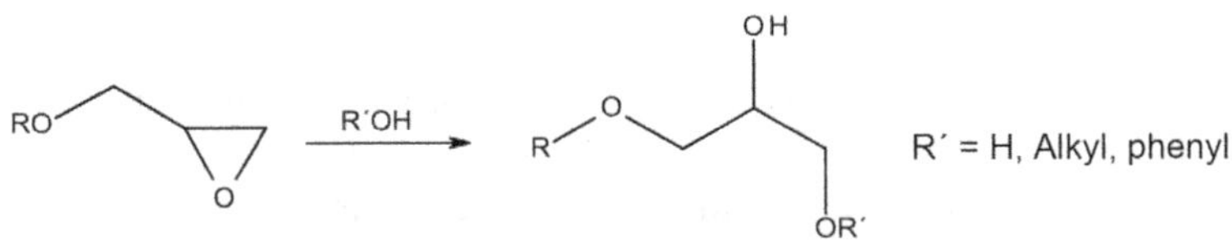

Equation 2.11: Reaction between OH groups and epoxides

The reaction with alcohols can be controlled with suitable bases. The reactivity of various alcohols with phenyl glycidyl ether has been shown to follow the order below (Equation 2.12) [19, 20].

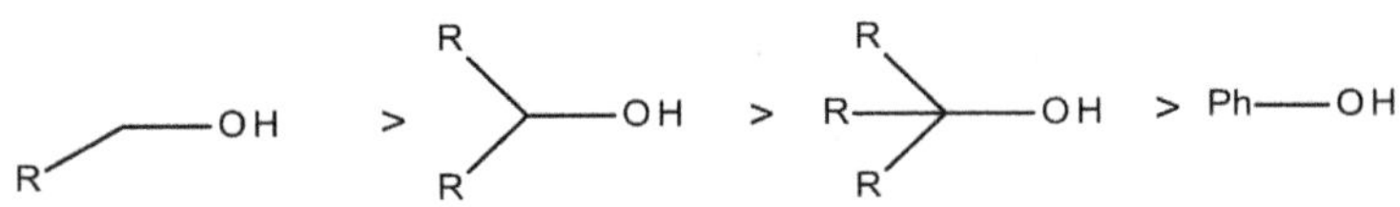

Equation 2.12: Order of reactivity of alcohol groups with epoxides [19, 20]

The reactivity of primary and secondary alcohols is particularly interesting because secondary OH groups are always generated during the reaction.

A large number of bases, including LiOH, $Mg(ClO_4)_2$, amines and other hydroxides, have been studied for their, in some cases, significantly varying selectivity [20].

The reaction with alcohols can be catalysed with aluminium oxide to such an extent that high conversion rates occur at 25 °C [17]. The products of this catalysis follow the regiochemistry of an S_N2 reaction. With cyclic compounds, only trans products are generated.

Reaction with compounds containing sulphur

Under base catalysis, hydrogen sulphide reacts with epoxy groups to yield ß-hydroxy thiols that normally react further to produce bis-(β-hydroxyalkyl)-sulphides [3].

Similarly, reaction with sulphites generates β-hydroxy sulphonates [3]. A high conversion rate can be achieved with a mixture of sodium hydrogen sulphite and sodium sulphite [16] and this reaction can be used to define the number of epoxy groups in a resin [15], although it must be remembered that organically bound chlorine also reacts (see Section 2.3).

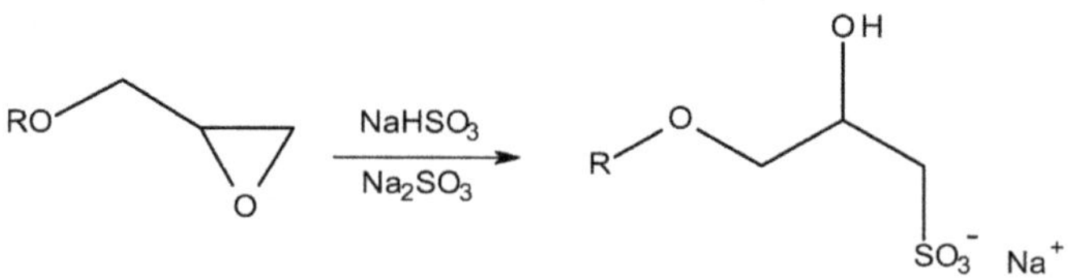

Equation 2.13: Reaction between sulphites and epoxy groups

Thiols react faster than amines with epoxides, especially at low temperatures, to form ß-hydroxy-sulphides in a reaction that can be accelerated by primary or secondary amines. The more alkaline the amine, the higher is the reaction rate [4].

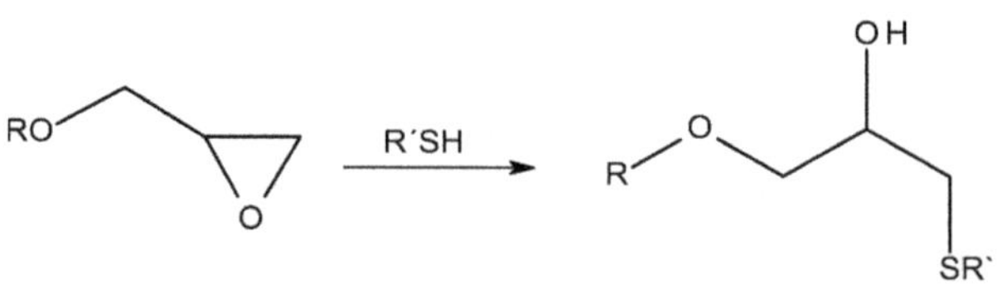

Equation 2.14: Reaction between thiols and epoxide groups

Isocyanates

The reaction between isocyanates and epoxides leads to oxazolidones [4]. For example, phenyl isocyanate yields N-phenyloxazolidone in a cycloaddition reaction which takes place at the N = C bond on the isocyanate [3].

Equation 2.15: Reaction between isocyanates and epoxides

Carbanion species

Epoxide groups react with malonic ester derivatives or acetoacetate esters to yield five-membered ring lactones [3].

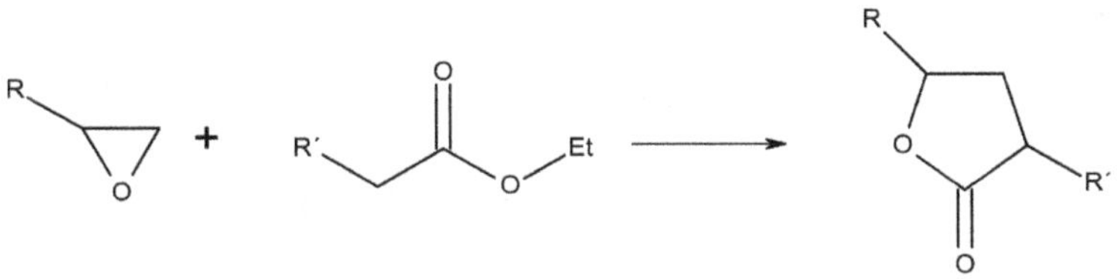

Equation 2.16: Reaction between carb-anion species and epoxides

2.1.2 Acid-catalysed reactions

Ring opening occurs when epoxy groups react with nucleophiles in the presence of acid catalysts. The reaction is similar to the cleavage of ether by HBr, but occurs in much milder conditions.

In acid conditions, ring opening follows an S_N2 reaction that has a great deal of S_N1 character [5, 7]. Other authors view this as a mixture of S_N2 and S_N1 reactions [1], although a clear S_N1 mechanism is observed in the case of benzylic substituents or of two alkyl substituents at a cyclic carbon atom [2], because formation of a carbenium ion is then stabilised by the electron-donating effect of the substituents.

An S_N1 reaction creates products with substitution at the more sterically hindered carbon atom, because that is where the carbenium ion can be stabilised most. In compounds which are based on epichlorohydrin and have only one alkyl substituent, the reaction proceeds via an oxonium ion [3] that can react according to either an S_N2 or an S_N1 mechanism.

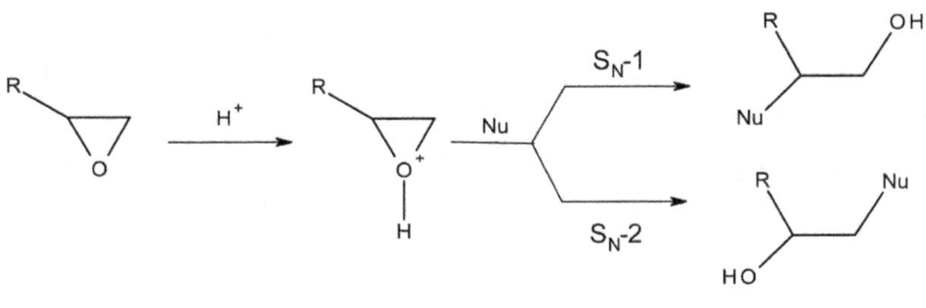

Equation 2.17: Mechanism behind acid catalysis of epoxide ring opening

As the two mechanisms take place concurrently, mixtures of both regioisomers are generated.

Reaction with hydrogen halides

Halohydrins are formed when epoxides react with hydrogen halides [3]. As this reaction is more or less quantitative with HBr, it is used to determine the number of epoxide groups in resins (see Section 2.3.1).

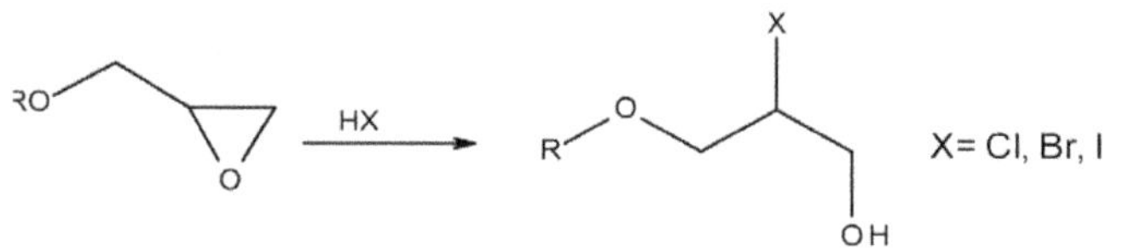

Equation 2.18: Reaction between epoxides and hydrogen halides [7]

Reaction with carboxylic acids

Epoxide groups react with carboxylic acids to produce β-hydroxy esters in very high yield in the presence of basic aluminium oxide and under mild conditions [3]. In the absence of a catalyst, high conversion rates require temperatures in excess of 150 °C. Ring opening leads to mixtures, with S_N1 products being preferentially formed in the case of glycidyls [7].

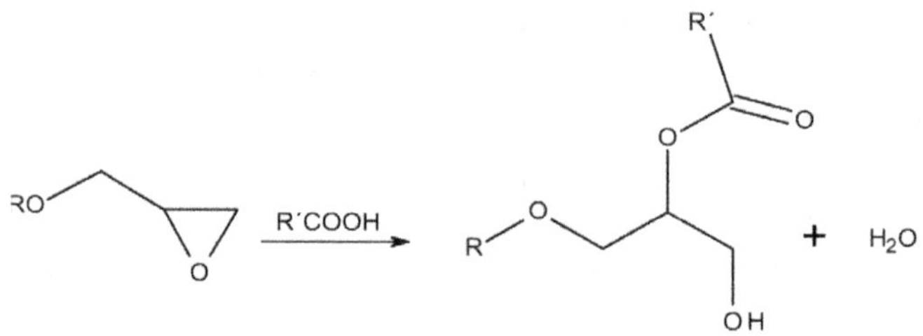

Equation 2.19: Reaction between carboxylic acids and epoxides

In [13], it is assumed that the products formed are the outcome of an S_N2 mechanism, with mixtures of both regioisomers being generated.

The resultant hydroxyl group can react with a carboxylic acid to yield an ester or with an epoxy group to create an ether, the generated water reacting with epoxy groups under ring opening to produce diols [4] (see Equation 2.22).

Reaction with phosphoric acid, phosphonic acid and derivatives

Epoxides react both with phosphoric acid and with phosphates to yield phosphoric acid esters, with the stoichiometry determining whether mono-, di- or tri-esters are produced [30–33].

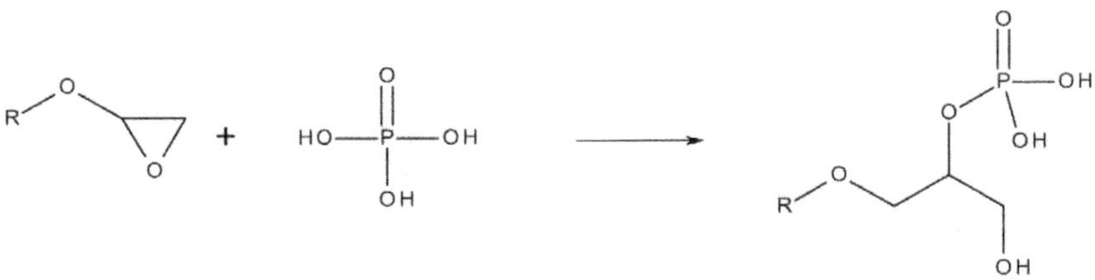

Equation 2.20: Reaction between phosphoric acid and epoxides [30]

According to patents [31–33], the same reaction should also be possible with phosphonic acids and phosphinic acids, but no examples can be found for these reactions. Phosphonic acids react with epoxy groups to yield cyclic phosphonic acid diesters [3].

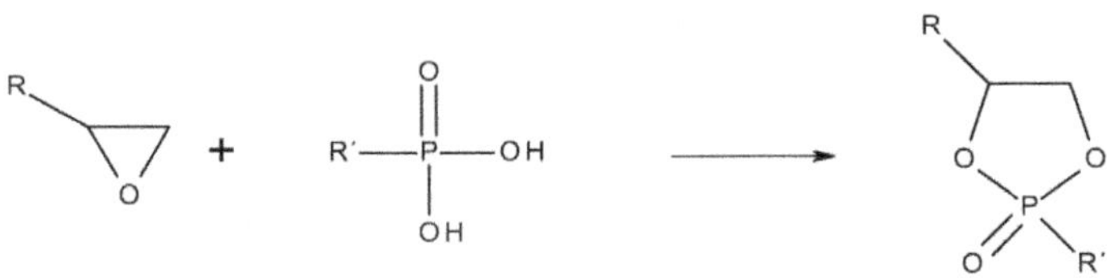

Equation 2.21: Reaction between phosphonic acids and epoxides

It is highly probable that the reaction pathway can be manipulated to preferentially yield the open mono-esters (see Section 3.5.1).

Reaction with water, alcohol and phenol

The reaction between epoxy groups and alcohols and phenols yields the same products under acid catalysis as under base catalysis, but the regioselectivity is different. Often, product mixtures are obtained [7].

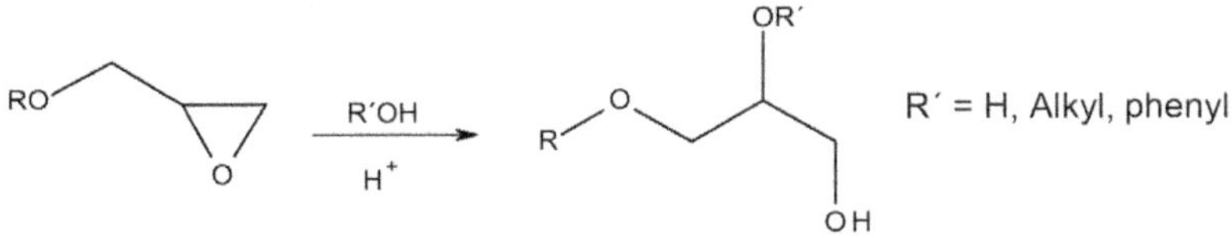

Equation 2.22: Acid-catalysed reaction between OH groups and epoxides

An overview of the most important acid and base catalyses is provided in Equations 2.23 and 2.24.

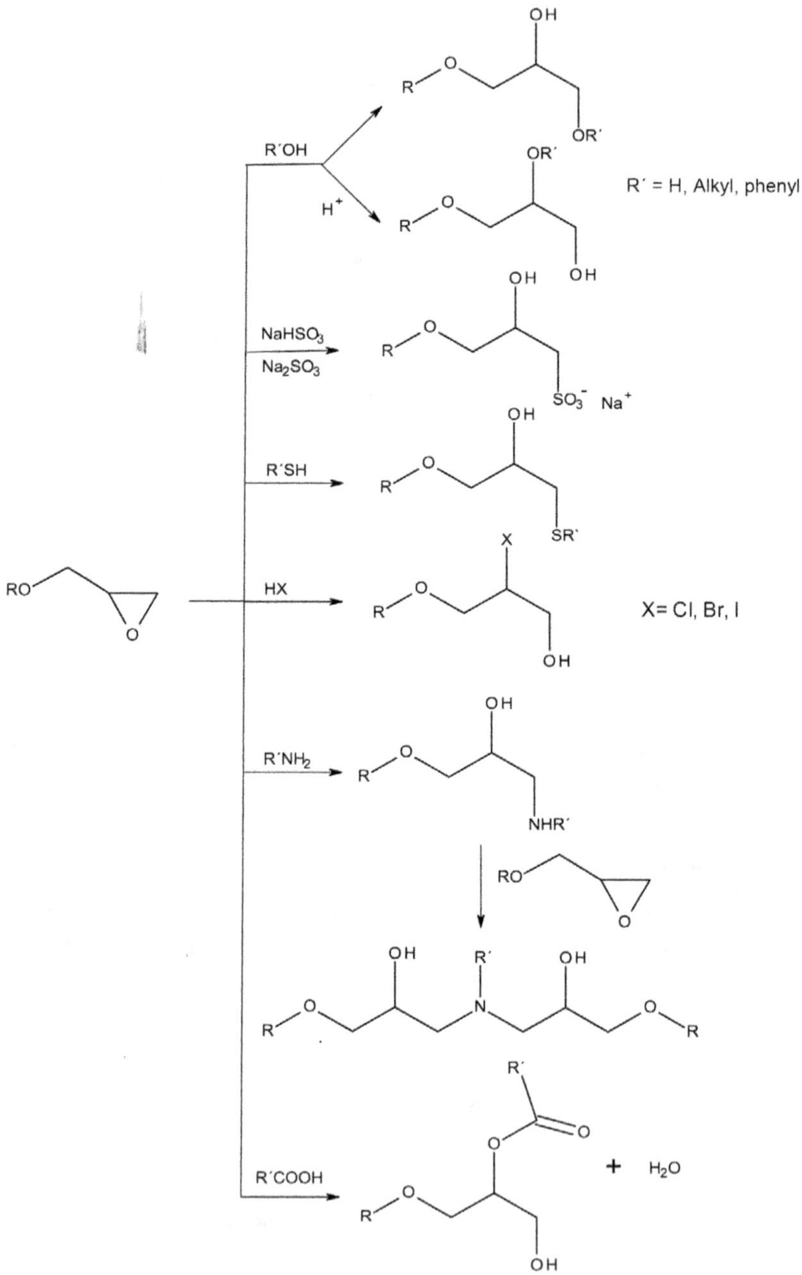

Equation 2.23: Reactions of epoxy groups I

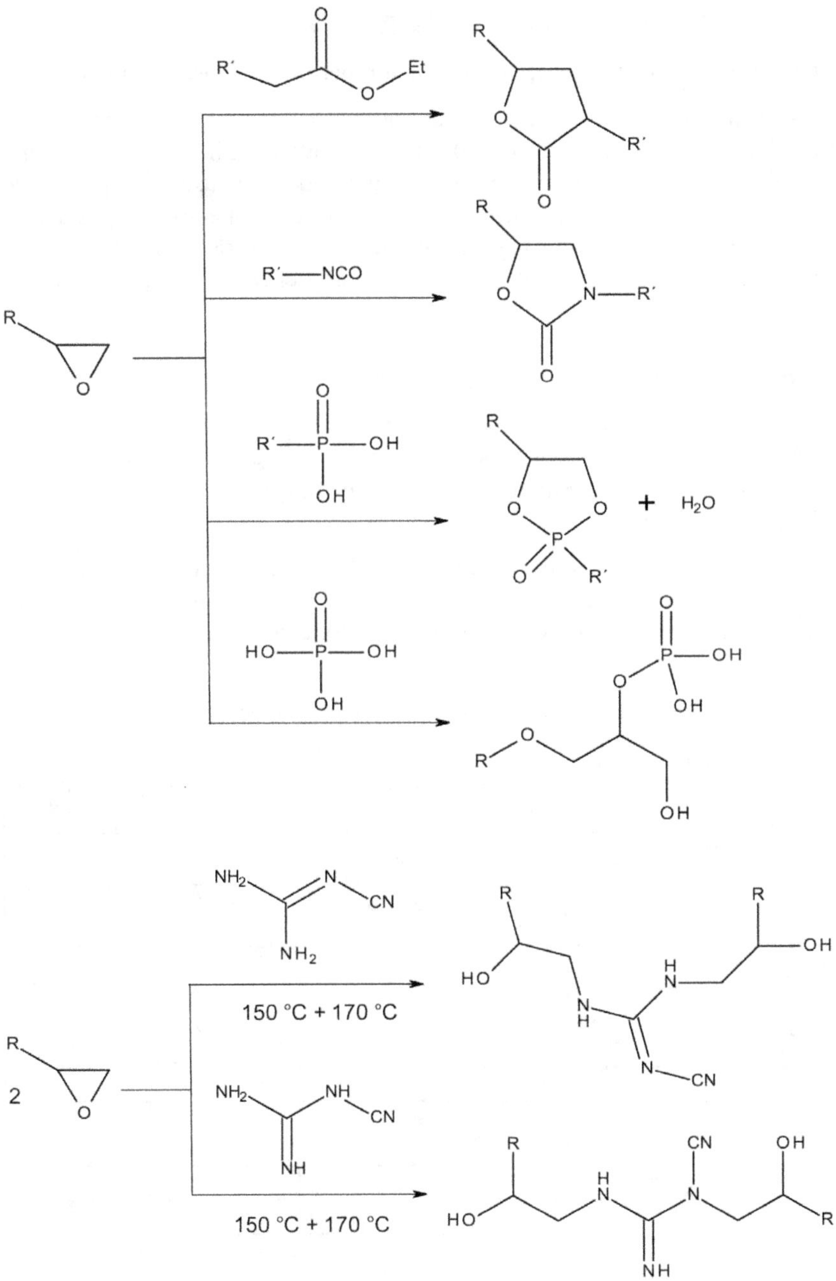

Equation 2.24: Reactions of epoxy groups II

2.1.3 Properties of the epoxy group

The epoxy group can be characterised by a number of spectroscopic methods.

Surface-sensitive methods are of particular interest to coatings technology, which is why infrared spectroscopy, especially in combination with ATR, has become well established in the coatings industry. This method can be used to characterise both liquid coatings and solid coatings on surfaces. NMR spectroscopy is useful for characterising polymers in solution and provides structural information about the compounds. Other important spectroscopic and chemical properties of the epoxides will also be presented below.

Infrared spectroscopy of the epoxy group

Infrared spectroscopy is ideal for analysing epoxides because the ring strain shifts the vibration frequency and any substituents on the ring greatly influence the vibration frequencies in the medium NIR range (see Section 2.2.2 for information on NIR spectroscopy) [1].

Small rings show a shift in the asymmetric C-H stretching band to higher wave numbers (3080 to 3000 cm^{-1}), with the wave number decreasing with increase in ring size, as can be seen in the series oxirane (3052 cm^{-1}), oxetane (2978 cm^{-1})

Table 2.1: IR vibration frequencies of some oxiranes and oxetanes

Compound	Ring vibration [cm^{-1}]	C-H stretching [cm^{-1}]	CH_2 deformation [cm^{-1}]	CH_2- out of plane [cm^{-1}]
	1266 [1]	3079, 3063, 3016, 3005 [1]	1490, 1470 [1]	1153, 1120 [1]
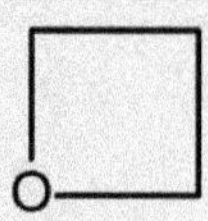	980-970, 900 [1]			
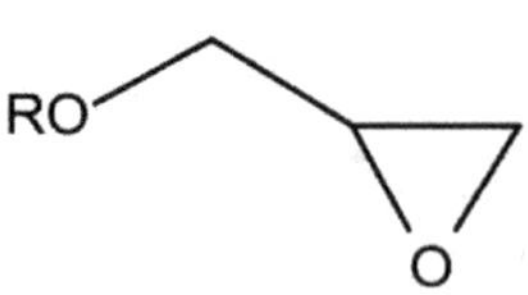	866 [26]	3050 [22]		
	915 [21,22, 25]	calculated [27] 3139, 3045		
	916 [18]			
	917 [23]			1190 [24]
	920 [24]			
	calculated [27] 987, 910, 855			

and tetrahydrofuran (2958 cm^{-1}) [1]. Similarly, ring vibration in oxirane occurs at higher wave numbers than in oxetane. Table 2.1 shows the most important vibration frequencies. Most authors state that the "epoxide band" of the glycidyl group (esters and ethers) occurs at roughly 920 cm^{-1}. Deviations from this can be explained in terms of the quantum mechanics of the vibration frequencies [27]. Due to the fact that many vibrations also contain contributions from the backbone (which is often BPA), three dominant vibrations occur, the intensities of which vary with the structure. Consequently, as a result of band coupling, the "epoxide band" may be shifted by up to 50 cm^{-1}. Instrument effects also play a role.

NMR spectroscopy of the epoxy group

NMR spectroscopy is ideal for analysing the molecular structure of the epoxides because ^{1}H-NMR and ^{13}C-NMR are capable of revealing changes in the backbone (Table 2.2).

Further properties of the epoxy group

Absorption by the oxirane group in the UV range is not suitable for characterising epoxides, as the longest wavelength absorbed by oxirane occurs at 171 nm [1] and is therefore beyond the reach of common spectrometers (lower limit at 210 nm).

In mass spectroscopy, epoxides undergo ring opening to yield a radical and a cation [1].

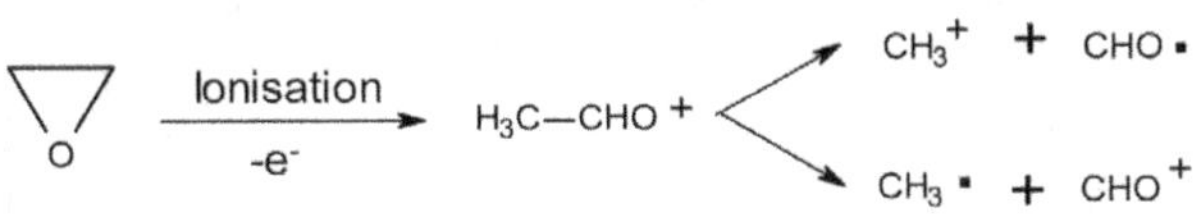

Equation 2.25: Fragmentation of epoxides in mass spectroscopy

Also possible is α-cleavage with retention of the ring [1], as shown in Equation 2.26.

Equation 2.26: a-Cleavage of epoxides in mass spectroscopy

Saturated heterocyclic compounds have similar alkalinity to open-chain compounds, apart from heterocyclic three-membered rings, which exhibit significantly reduced alkalinity. The proton affinity of gaseous ethylene oxide is 793.3 kJ/mol and therefore much less than that of dimethyl ether, at 807.9 kJ/mol [1].

Table 2.2: NMR data for some three-membered rings. The emboldened numbers correspond to the atoms numbered in the structures.

Compound	^{1}H-NMR [ppm]	^{2}J [Hz]	cis ^{3}J [Hz]	trans ^{3}J [Hz]	^{13}C-NMR [ppm]
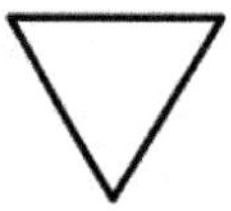	0.198 [63] 0.22 [1, 62]	-4.34 [63] -4.5 [62] -3 to -1 [1]	8.9 [63] 6 to 12 [1]	5.58 [63] -4–8 [1]	-2.2 [1] -2.8 [62]
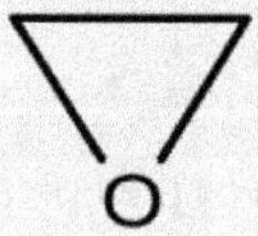	2.54 [64] 2.58 [62]	5.5 [62] 5 to 7 [1]	4.45 [64] 2 to 5 [1]	3.1 [64] 1 to 3 [1]	39.7 [1] 39.5 [62]
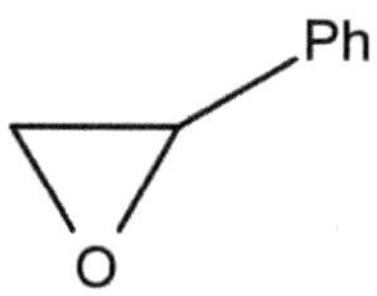	Multipletts: 2.8; 3.2; 3.9 [62]	5.5 [62]	4.1 [62]	2.6 [62]	
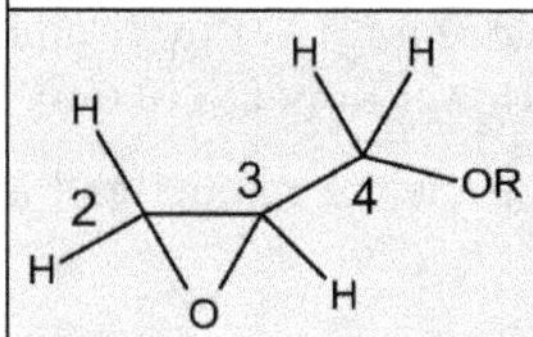	2.8 [21] 2.5–2.9; 3.2–3.5 [24, 28] **2**: 2.5–2.9; **3**: 3.1–3.5; **4**: 4.0–4.1 [24] **2**: 2.69–2.82 [69]; **3**: 3.28 [69]; **4**: 3.87–4.27 [69]	**2**: 5.15 [69] **4**: 11.5 [69]	**2**: 4.25 [69] **3**: 6.35 [69]	**2**: 2.6 [69] **3**: 2.85 [69]	44; 50 [21, 26] 2: 44.6; 3: 50.1; 4: 68.7 [65] 2: 44; 3: 50; 4: 68 [26] **2**: 44,2; **3**: 50.3; 69.9 [29]
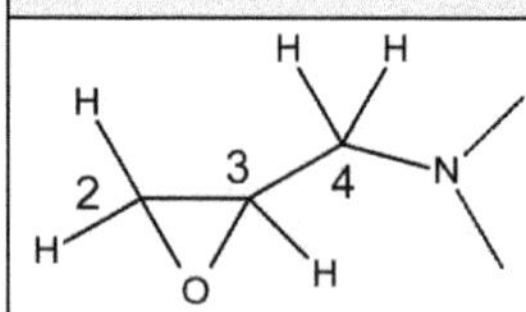	**2**: 2.6–2.93; **3**: 3.12–3.40; **4**: 3.72–3.82 [66]				

In 1929, *Brönsted* [34] observed that an aqueous KCl solution has an alkaline pH after addition of an epoxide. *Buddrus* rediscovered this effect in 1972 and used it in several synthetic routes [34–37].

Contact between the halide and the epoxy group, an equilibrium is established between the epoxy group and the 1,2-alkyl halide alcohol, which is generated via a 1,2-halide alcoholate intermediate; see Equation 2.27.

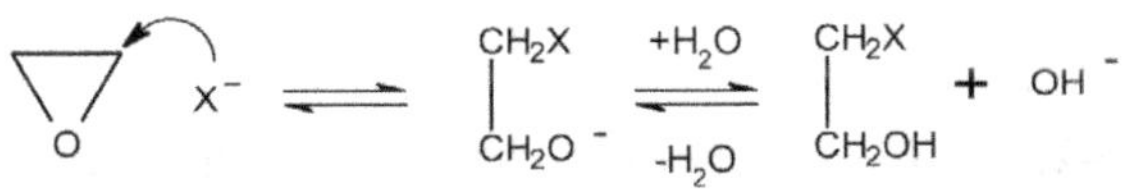

Equation 2.27: Equilibrium established after the reaction between an epoxide and a halide

The equilibrium leans heavily in favour of the epoxy group, so that the concentration of the alcohol in the mixture of ethylene oxide and chloride is about 2 % [34]. However, the low concentration is still sufficient to catalyse some reactions. For instance, the 1,2-halide alcoholate is sufficiently alkaline to deprotonate phosphonium salts, which then react with carbonyl groups to yield olefins [34, 36, 37], as shown in Equation 2.28.

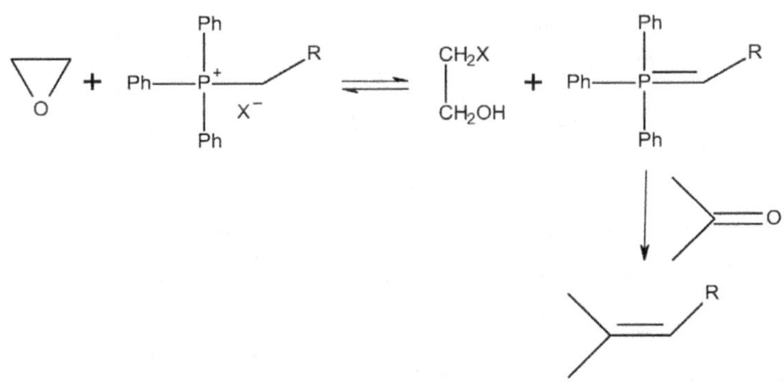

Equation 2.28: Reaction between an alkaline epoxide/halide mixture and a phosphonium salt

This alkaline property in connection with halides, which are always present in tiny concentrations, may be the cause of several side reactions of epoxides. The number of such side reactions can be illustrated by reference to the reactions between epoxides and metal salts. Damp epoxides (ethylene oxide) react with calcium chloride to form calcium hydroxide and the corresponding glycol chlorohydrin, and zinc chloride reacts with ethylene oxide over several months to yield polymers of ethylene oxide of different chain lengths (see Section 2.4.2) [38]. Finally, the reaction in which iron(III) chloride and epichlorohydrin generate 1,3-dichloropropanol and iron oxide hydrate is also mentioned in the literature [38].

2.2 Production and properties of epoxy resins

2.2.1 Production and properties of the monomers

2.2.1.1 Epichlorohydrin

Chloromethyl oxirane, 1-chloro-2,3-epoxy propane or epichlorohydrin is produced by chlorinating propene to allyl chloride, followed by reaction with hypochlorous acid to the dichlorohydrin of glycerol, and then dehydrochlorination to epichlorohydrin through reaction with sodium hydroxide [4], as shown in Equation 2.29.

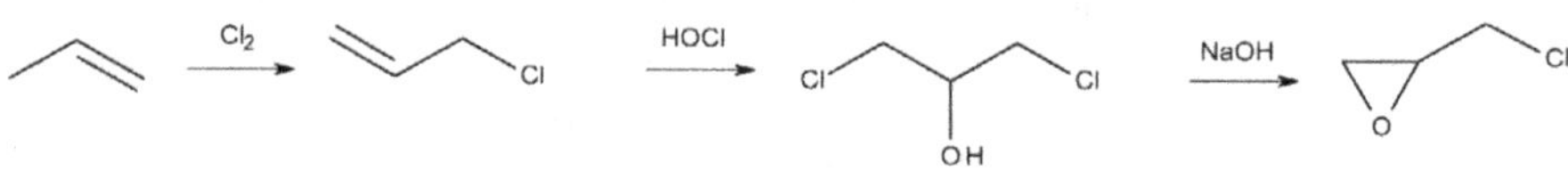

Equation 2.29: Production of epichlorohydrin

In industry, production is a two-stage reaction between allyl chloride and chlorine and water in the presence of milk of lime. First, the allyl chloride is made to react with HOCl at 25 to 30 °C (30 to 55 °C [6]) to yield a mixture of both dichlorohydroxy propanes, followed by reaction in the second stage with milk of lime at 50 to 90 °C (60 to 100 °C [6]) to yield epichlorohydrin and CaCl2 [39]. The halonium ion attacks at the less sterically hindered side of the double bond and ring formation happens on the opposite side, i.e. the sterically hindered side generates a highly reactive epoxy group. The mechanism behind the formation of epichlorohydrin explains its high reactivity, especially compared with the less reactive cycloaliphatic epoxides produced with peroxy acids [7].

Table 2.3: Properties of epichlorohydrin

Properties	Epichlorhydrin
General	Colourless liquid with chloroform like odor
Melting point [°C]	-48
Boiling point [°C]	116.56
n_D^{20}	1.4382 [3], 1.437–1,439 [6]
Good solubility in	Alcohols, ester, ether, ketones, aromates
Solubility in water [%]	6.6 [3], 7 [6]

As production of biodiesel continues to rise, the quantity of crude glycerol is increasing, too. Every 9 kg of biodiesel produced generates 1 kg of crude glycerol solution as by-product with a content of 15 % pure glycerol [40]. This renewable material can also be used to produce epichlorohydrin. In the Solvay process [41], glycerol is chlorinated with HCl and a catalyst [42] to yield dichlorohydroxy propane, followed by reaction with sodium hydroxide solution to give epichlorohydrin [40].

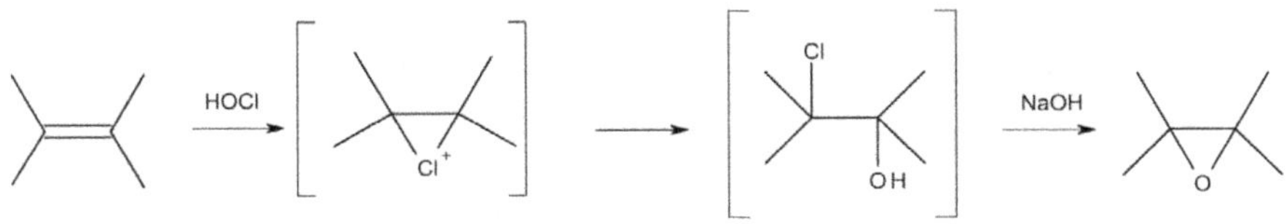

Equation 2.30: Mechanism of the epoxidation with HOCl

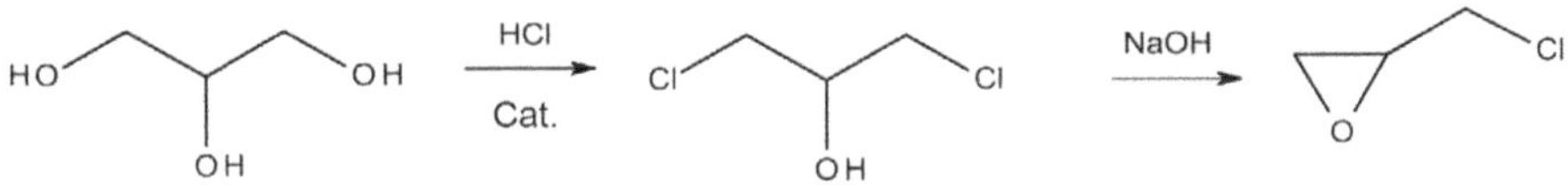

Equation 2.31: Synthesis of epichlorohydrin from glycerol

2.2.1.2 Bisphenols

Bisphenol A (BPA) is produced by the reaction between two moles of phenol and one mole of acetone [4].

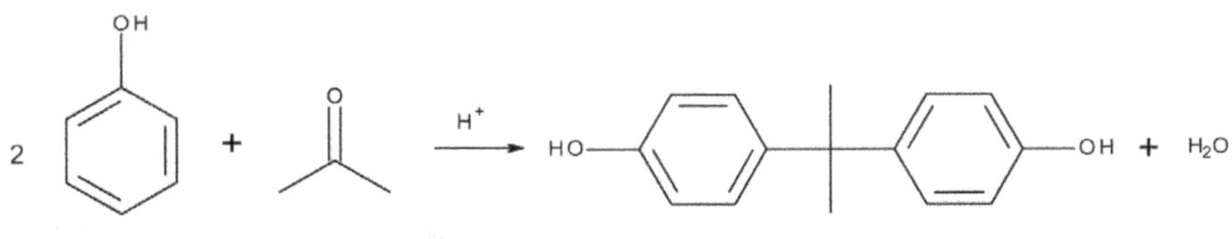

Equation 2.32: Production of bisphenol A

In general, bisphenols can be produced by condensing phenol with aldehydes and ketones [6].

Bisphenols based on bisphenol F (BPF) are produced via a formaldehyde coupling, which generates a mixture of isomers [4].

Similarly, a tetra-phenol is obtained when glyoxal is used [4].

The other bisphenols are produced by the reaction of ketones and phenol in acidic solution, with even sterically hindered ketones, such as cyclohexanone, yielding bisphenol Z (BPZ) in high yields. Highly alkylated bisphenols are produced by downstream alkylation of the bisphenol [45].

Similarly, bisphenols that contain carboxyl groups can be obtained from ketocarboxylic acids [6]. One such example, is diphenolic acid, which is produced by making levulinic acid (4-oxopentane acid) react with bisphenol A [47, 48].

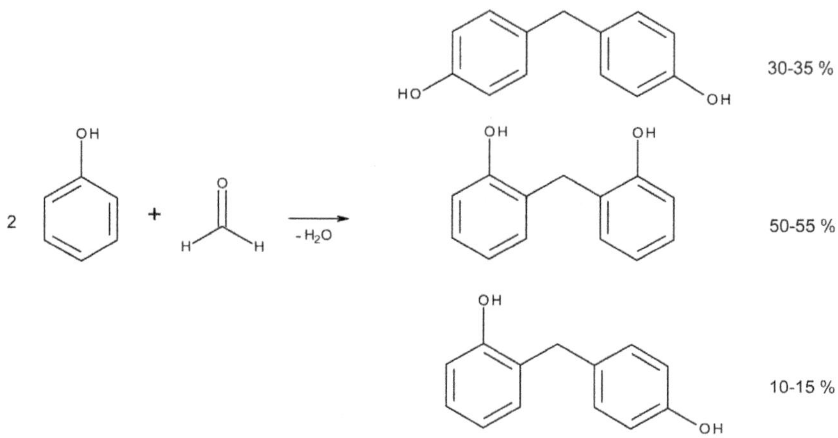

Equation 2.33: Isomeric composition in the production of bisphenol F

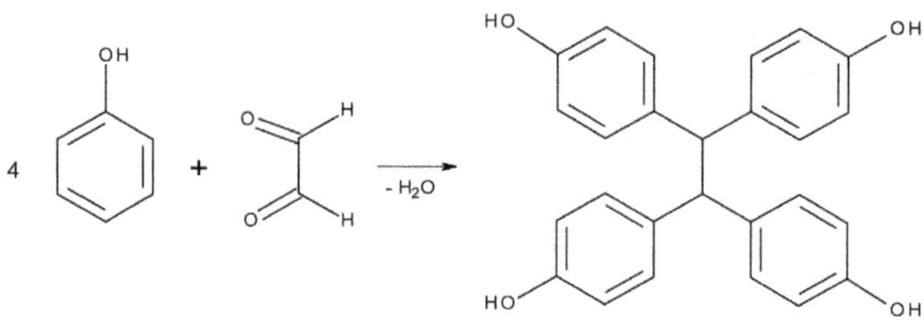

Equation 2.34: Production of tetrakis(4-hydroxyphenyl)-ethane from glyoxal and phenol

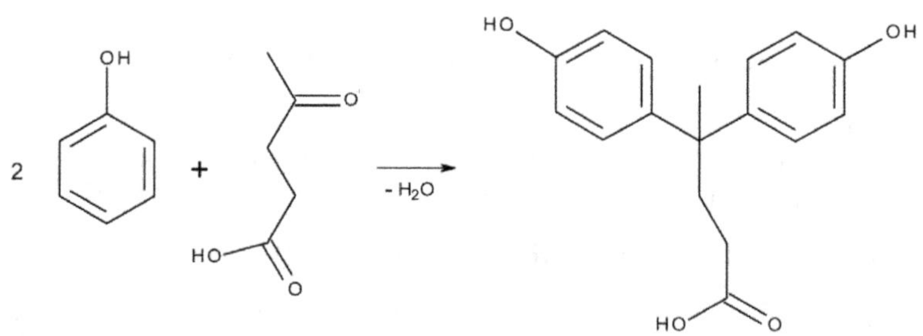

Equation 2.35: Production of diphenolic acid

As levulinic acid can be produced from hydroxymethylfurfural, which in turn can be produced from carbohydrates [47, 48], and as phenol can be generated from lignin [49], diphenolic acid thus constitutes a renewable material for epoxy resins. Table 2.4 lists some important bisphenols and their properties [45].

Table 2.4: Important bisphenols and their properties, according to [45]

Name	Trivial name	Structure	Totals formula	Molecular weight [g/mol]	Melting point [°C]	Boiling point [°C]
Bis-(4-hydroxy-phenyl)-methane	Bisphenol F	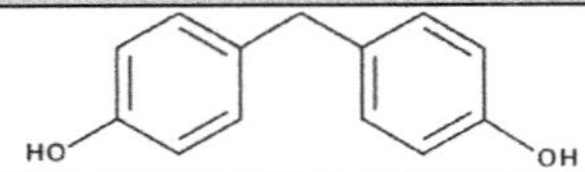	$C_{13}H_{12}O_2$	200.2	158	
2,2-Bis-(4-hydroxy-phenyl)-propane	Bisphenol A	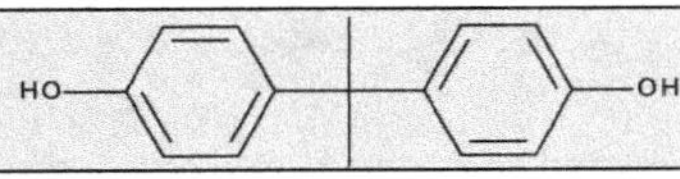	$C_{15}H_{16}O_2$	228.3	156-157	190 (1.3 mbar)
4,4-Bis-(4-hydroxyphenyl)-valeric acid	Diphenylic acid	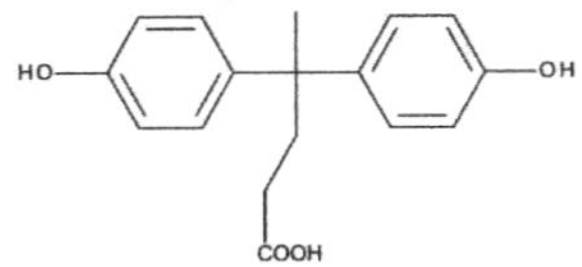	$C_{17}H_{18}O_4$	286.3	173	
2,2-Bis-(3-methyl-4-hydroxyphenyl)-propane	Bisphenol C [46]	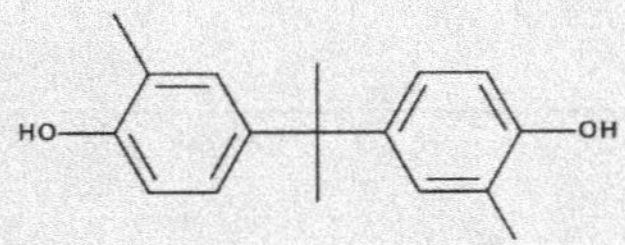	$C_{17}H_{20}O_2$	256.3	136	
2,2-Bis-(4-hydroxy-phenyl)-cyclohexane	Bisphenol Z	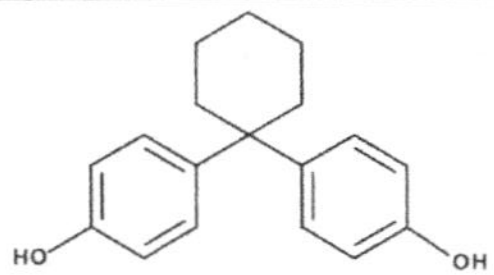	$C_{18}H_{20}O_2$	268.4	188	
Bis-(2-hydroxy-3-tert.-butyl-5-methyl-phenyl)-methane		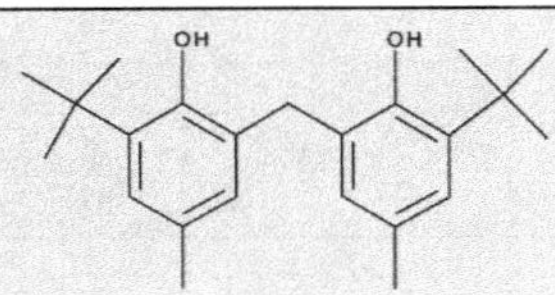	$C_{23}H_{32}O_2$	340.5	131	

Table 2.4 Continue

Name	Trivial name	Structure	Totals formula	Molecular weight [g/mol]	Melting point [°C]	Boiling point [°C]
a,a'-Bis-(4-hydroxyphenyl)-p-diisopropylbenzol		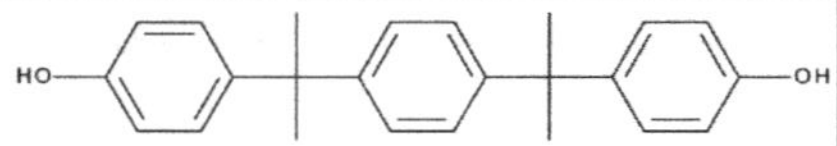	$C_{24}H_{26}O_2$	346.3	196	230 (0.13 mbar)
Bis-(2-hydroxy-3-cyclohexyl-5-methylphenyl)-methan		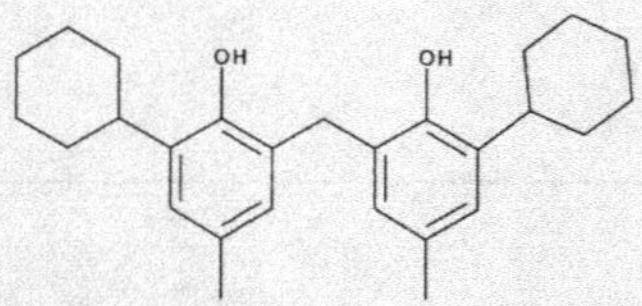	$C_{27}H_{36}O_2$	392.6	128	
Bis-(4-hydroxy-3,5-di-tert.butylphenyl)-methan		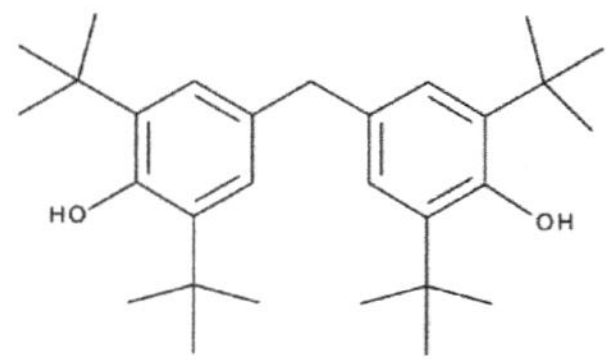	$C_{29}H_{44}O_2$	424.7	154	
1,1-Bis-(4-hydroxyphenyl)-ethan	Bisphenol E [46]	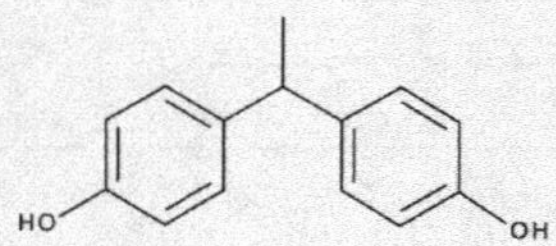	$C_{14}H_{14}O_2$	214.3		
2,2-Bis-(4-hydroxy-phenyl)-butan	Bisphenol B [46]	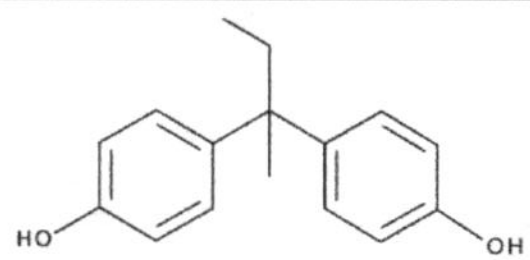	$C_{16}H_{18}O_2$	242.3		

The reaction between bisphenols and epichlorohydrin, followed by dehydrohalogenation, yields bifunctional epoxides, as illustrated by 2,2-bis[4-(2,3-epoxypropoxy)phenyl]propane (bisphenol A diglycidyl ether (DGEBA)) [24].

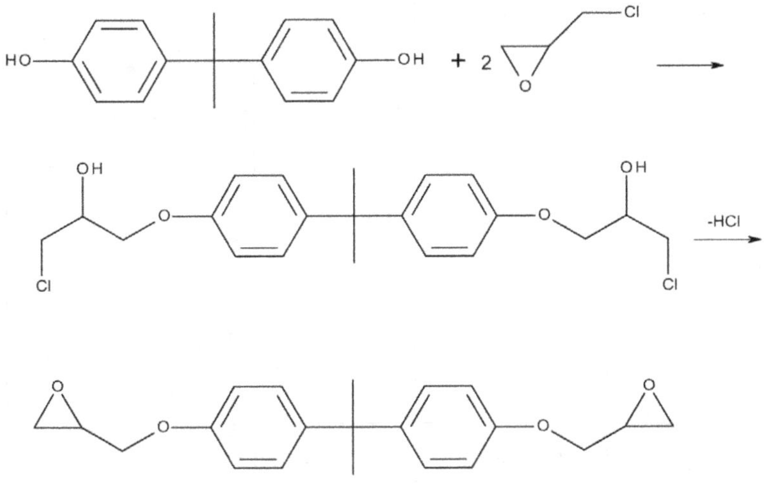

Equation 2.36: Production of bisphenol A diglycidyl ether

Mixtures of oligomers are usually generated (see Section 2.2.2) because the side reaction between diglycidyl ether DGEBA and bisphenol A yields hydroxy-functional diglycidyl ether (see Section 2.4.2).

At a 10 : 1 ratio of epichlorohydrin to bisphenol A, the monomer content in the product mixture is 90 % [24]. These mixtures are called low-molecular resins [24]; the commercially available products contain 80 % DGEBA [4]. The effect of the mixing ratios of epichlorohydrin to bisphenol A is shown in Table 2.5.

Besides the aforementioned bisphenols, a hydrogenated bisphenol A diglycidyl ether has been commercially available since 1976, and is produced by reaction of the corresponding alcohol with epichlorohydrin or by hydrogenation of the bisphenol A diglycidyl ether or its oligomers [3].

Highly-functional epoxides, such as 1,1,2,2-tetrakis[4-(2,3-epoxypropoxy) phenyl]-ethane are added to epoxide formulations to improve the chemical resistance of the resultant coating [24].

Table 2.5: Effect of mass ratio on molecular weight [4]

Epichlorhydrin: bisphenol A	Molecular weight [g/mol]
10:1	370
2:1	450
1.4:1	791

Table 2.6: Properties of important bisphenol diglycidyl ethers

Name	Structure	Total formula	Molecular weight [g/mol]	Viscosity at 25 °C [mPas]	EEW* [g]
Bisphenol-A-diglycidylether	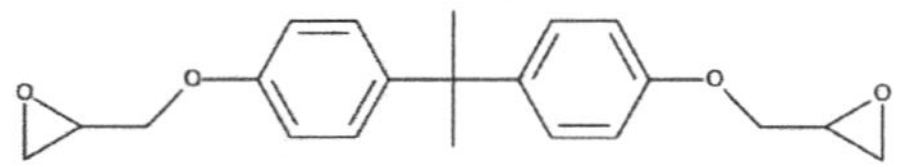	$C_{21}H_{24}O_4$	340.4	8000–15000 [7]	180–190 [7], 175–200 [24]
Bisphenol-F-diglycidylether	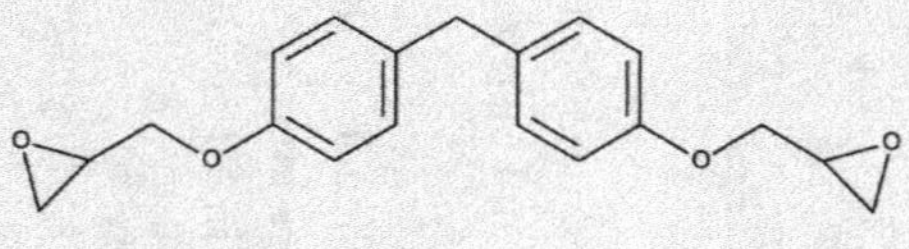	$C_{19}H_{20}O_4$	312.4	3000–5000 [7]	170–180 [7]
2,2-Bis[4-(2,3-epoxypropoxy)cyclohexyl] propane	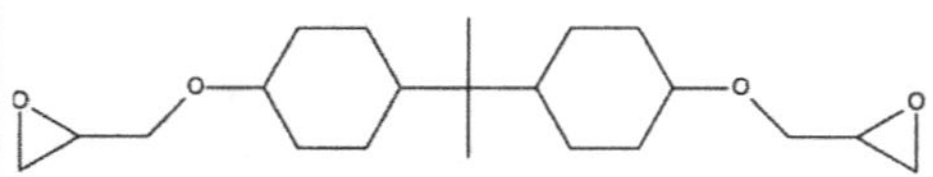	$C_{21}H_{36}O_4$	352.5	2000–2500 [3, 24]	232–238 [3, 24]
1,1,2,2-Tetrakis[4-(2,3-epoxypropoxy)phenyl] ethane	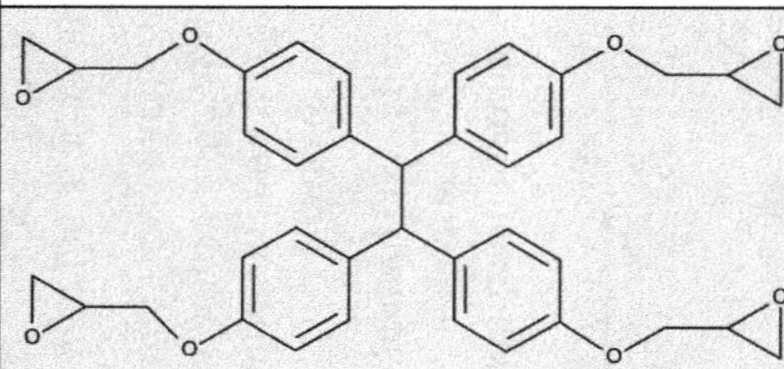	$C_{39}H_{42}O_8$	638.7		185–208 [24]

2.2.1.3 Epoxides based on olefins

Several aliphatic epoxy resins can be obtained by direct epoxidation of an olefin with a peroxy acid [4].

Equation 2.37: Epoxidation of double bonds with a peroxy acid

The *Prilezhaev* reaction proceeds via the mechanism proposed by *Bartlett*, which is known as the butterfly mechanism [43, 44] because of the geometry of the cyclic transition state [3].

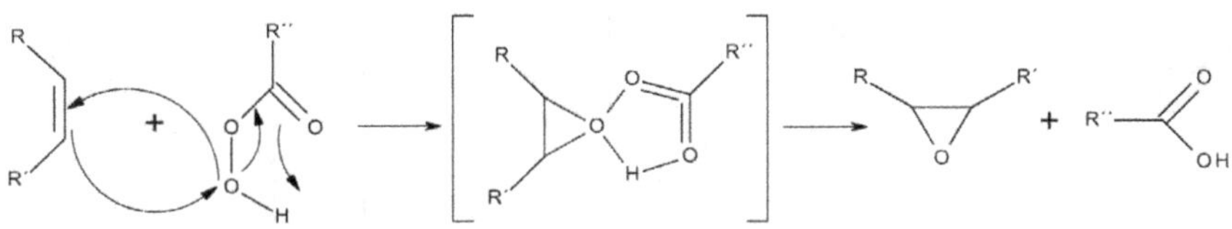

Equation 2.38: Mechanism of epoxidation with a peroxy acid, according to Bartlett (Buttterfly mechanism)

The epoxidation is regio-selective and insertion of the oxygen occurs on the less sterically hindered side of the double bond [7]. The resulting epoxides are therefore less reactive than those formed with epichlorohydrin.

Single-molecule fluorescence measurements [97] have shown that, besides the concerted mechanism, another mechanism happens which proceeds via a protonated epoxy ring as a transition state.

The epoxidation is a second order reaction and highly exothermic (250 kJ/mol). It therefore requires very close control [6]. Electron-donating groups, such as alkyl, accelerate the reaction at the double bond whereas electron-withdrawing groups have the opposite effect. Cyclic olefins react faster than dialkyl-substituted open-chain compounds, but phenyl substituents have only a minor accelerating effect [3, 6]. The advantage of producing epoxides in this way is that it generates non-halogen products, which are in great demand in the electronics industry [6]. Furthermore, side reactions arising from the alkaline effect of halides in conjunction with epoxy groups can be ruled out. In addition, cycloaliphatic epoxides have a low viscosity, high UV stability and good comparative tracking index [3, 7]. A number of cycloaliphatic epoxides are presented in Table 2.7.

Table 2.7: Properties of some cycloaliphatic epoxides

Name	Structure	Total formula	Molecular weight [g/mol]	Viscosity at 25 °C [mPas]	EEW* [g]
2-(3,4-Epoxycyclohexyl)-oxirane	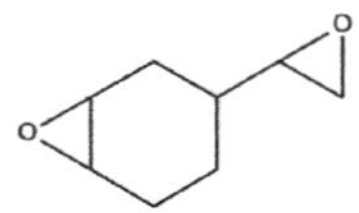	$C_8H_{12}O_2$	140.2	15 [3, 4], 7 [6]	76 [3], 70–74 [4]
3',4'-Epoxycyclohexyl-methyl-3,4-epoxy-cyclohexancarboxylate	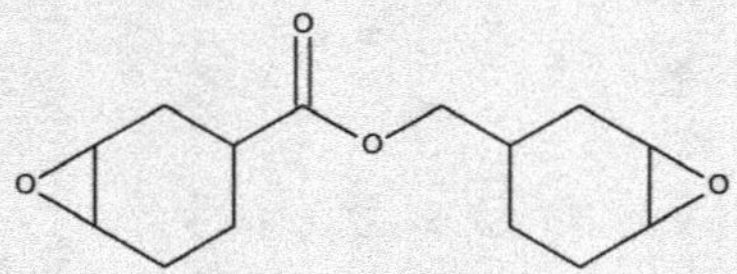	$C_{14}H_{20}O_4$	252.3	300–400 [7], 350 [3], 350–450 [4]	135–145 [7], 135 [3], 131–143 [4]
Bis-(3,4-epoxycyclohexyl-methyl)adipic ester	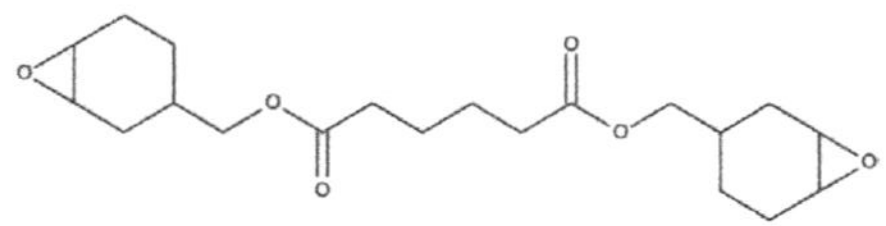	$C_{20}H_{30}O_6$	366.5	550–750 [4]	180–210 [4]
2-(3',4'-epoxycyclohexyl)-5,1''-spiro-3'',4''epoxy-cyclohexane-1,3-dioxane	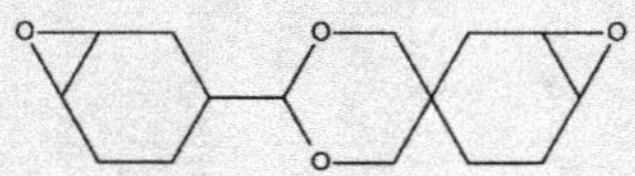	$C_{15}H_{22}O_4$	266.3	7000–17000 [4]	133–154 [4]

* See Section 2.3 for a definition of EEW

2.2.1.4 Glycidyl esters

Glycidyl esters are produced in the same manner as bisphenolglycidyl ether by reaction between mono- or di-carboxylic acids and epichlorohydrin, followed by dehydrohalogenation [4].

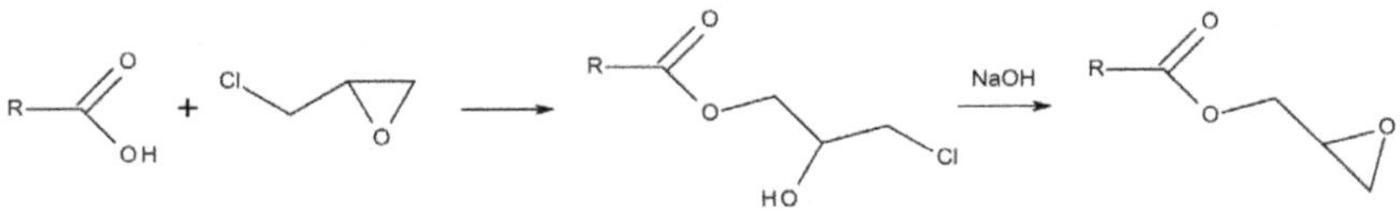

Equation 2.39: Production of glycidyl esters from carboxylic acids and epichlorohydrin

These esters usually have a very low viscosity of around 500 mPa s, but they have similar reactivity to bisphenol A-based epoxy resins [4].

Hexahydrophthalic diglycidyl ester is notable for its high arc resistance and comparative tracking index and is therefore used in insulating compounds [6].

A special group within this class of compounds is made up of the glycidyl esters of versatic acid, which impart outstanding chemical resistance to coatings and are therefore used in clearcoats and topcoats. Their high stability stems from extensive steric hindrance of the carboxylic group by the three, usually long-chain alkyl groups. These kinds of carboxylic groups can therefore only be esterified in concentrated sulphuric acid via an acylium ion by an AAC1 mechanism [52]. This mechanism cannot take effect in dilute acids and alkaline solutions and so esterification cannot occur in them either.

2.2.1.5 Aliphatic glycidyl ethers

Apart from the glycidyl ether of cresol, these ethers are based on aliphatic alcohols. Owing to their low viscosity, they serve as reactive diluents or plasticisers [4, 6]. Commercial products include butan-1,4-diol diglycidyl ether, hexan-1,6-diol-diglycidyl ether, polypropylene glycol-diglycidyl ether and glycerol-diglycidyl ether [6].

Aliphatic monoglycidyl ethers also act as reactive diluents, e.g. 10 % addition of butylglycidyl ether to an epoxy resin lowers the viscosity by factor 10 [53].

2.2.2 Production and properties of the oligomers

2.2.2.1 Bisphenol-based epoxy resins

Epoxy resins based on bisphenol A are categorised by molecular weight, number of epoxy groups, number of OH groups, melting point and viscosity. Table 2.10 presents an overview of the commonest such resins. This classification is not rigid and has been modified repeatedly over the last 40 years, even though the resins have barely changed during that period.

Table 2.8: Properties of some glycidyl esters

Name	Structure	Total formula	Molecular weight [g/mol]	Viscosity at 25 °C [mPas]	EEW* [g]
Hexahydrophthalic acid-diglycidylester	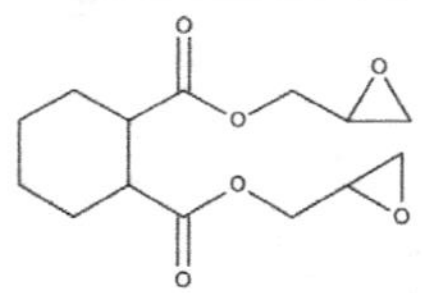	$C_{14}H_{20}O_6$	284.3	400–600 [7]	170–180 [7]
Neodecan-glycidyl ester	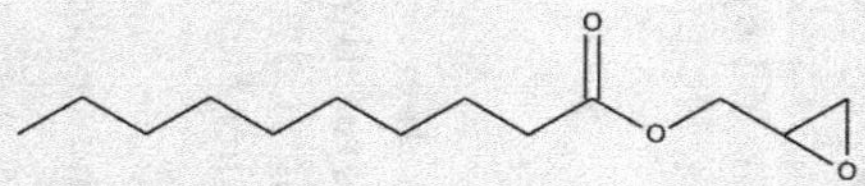	$C_{13}H_{24}O_3$	228.3	5–15 [7]	245–255 [7]
Acrylic acid-glycidyl ester	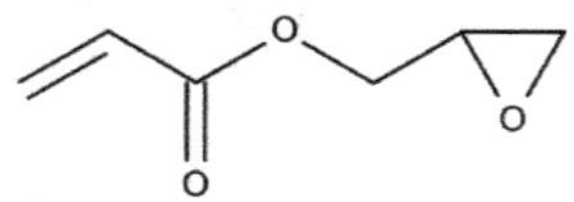	$C_6H_8O_3$	128.1	2.5 (20 °C) [6]	128
Methacrylsäure-glycidyl ester	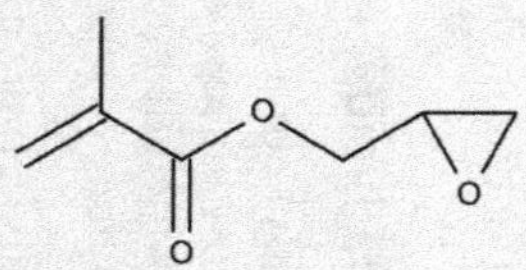	$C_7H_{10}O_3$	142.2		142
Versatic acid-glycidylester [50]	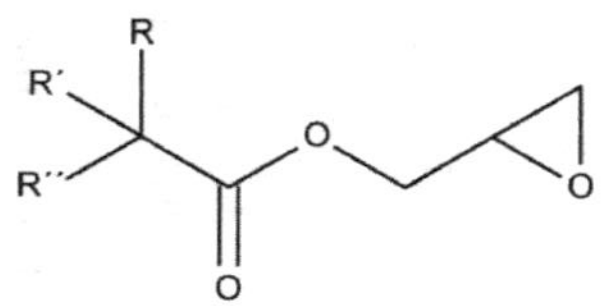	$C_{12}H_{22}O_3$ – $C_{14}H_{26}O_3$		7 [51]	240–250 [50]

* For a definition of EEW, see Section 2.3

Table 2.9: Properties of some glycidyl ethers

Name	Structure	Total formula	Molecular weight [g/mol]	Viscosity at 25 °C [mPas]	EEW* [g]
Butylglycidyl ether	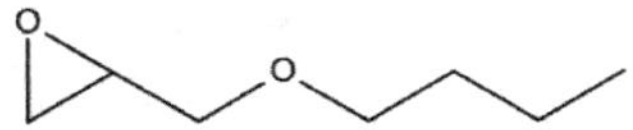	$C_7H_{14}O_2$	130.2	1–3 [53]	130–159 [53]
1,4-bis(2,3-epoxy-pro-poxy)butane	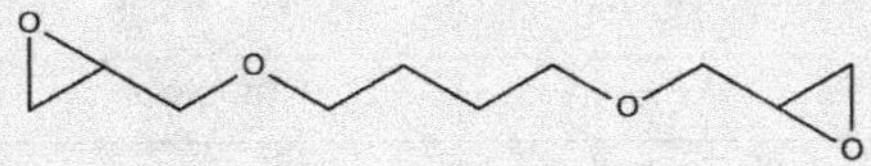	$C_{10}H_{18}O_4$	202.2	12–18 [3]	125–135 [3]
Hexandiglycidyl ether	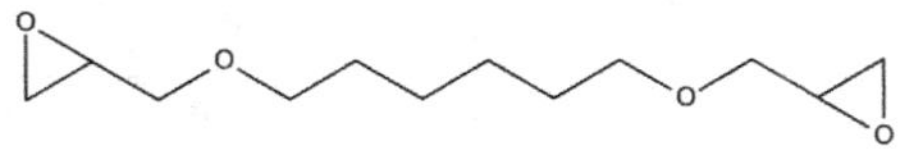	$C_{12}H_{22}O_4$	230.3	15–25 [7]	135–150 [7]
C12-C14-Alkylglycidyl ether	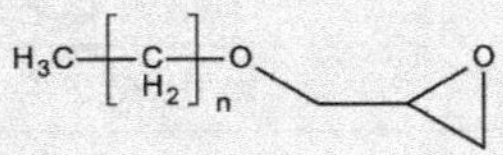 n=11–13			8–12 [7]	290–310 [7]
o-Cresylglycidyl ether	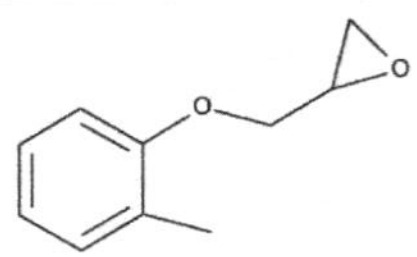	$C_{10}H_{12}O_2$	164.2	8–12 [7]	175–185 [7]
Polyalkylenoxid-glycidyl ether	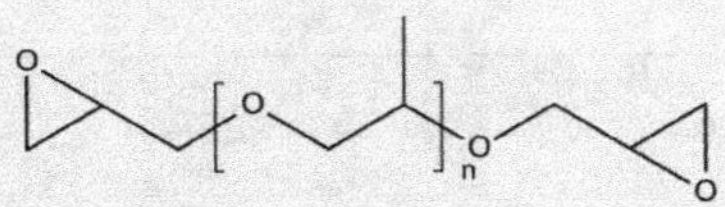 n=3–10			50–100 [7], 55–100 [3]	300–400 [7], 305–335 [3]

* For a definition of EEW, see Section 2.3

Table 2.10: Properties of bisphenol A resins

Property	Bisphenol A resins				
	Low molecular weight (liquid epoxy resin)	Middle molecular weight	Middle molecular weight	High molecular weight	Very high molecular weight
EEW * [g]	182–192 [24, 67]	500–560 [24, 67]	875–950 [67]	1600–2000 [24], 1600–2300 [67]	2500–5500 [67]
	188 [4]	450–525 [71, 72]	870–1025 [71,72]	1650–2050 [71,72]	2400–4000 [71, 72]
	225–290 [71]	450–500 [73]	850–940 [73]	1700–2050 [73]	2500–4000 [73]
	175–290 [72]				
Epoxy groups/Mol [72]	1.8–2	1.8	1.54	1.45	1.15
Repeating units [n]	0.06–0.14 [24]	2.3–2.7 [24]	5.5 [67]	10–13 [24]	>16 [67]
	0.11–0.15 [67]	2 [67, 71]	3.7 [71]	14.4 [67]	
	0.5 [71]			8.8 [71]	12 [71]
Viscosity of the molten mass [Pa*s]	11.0 at 14.5 at 25 °C [4, 24]	0.4–0.8 at 150 °C [24]		30–100 at 150 °C [24]	
Melting point [°C]	liquid at RT	65–85 [67]	90–110 [67]	120–135 [67]	130–160 [67]
		64–76 [71]	95–105 [71]	125–132 [71]	140–155 [71]
Hydroxyl groups/Mol	0.75 [71]	2.8 [71, 72]	4.8 [71, 72]	10.5 [71, 72]	15.5 [71, 72]
	0.3–0.75 [72]				
Hydroxyl value **	0.07–0.16 [72]	0.32 [72]	0.34 [72, 73]	0.36 [72, 73]	0.40 [72, 73]
		0.29 [73]			

* See Section 2.3 for a definition of EEW and OH value, ** calculated with the assumption of two epoxy groups in the chain

There are two established synthetic routes for epoxy resins based on BPA. The first uses epichlorohydrin and bisphenol A as reactants and is variously called the **Taffy process** [4] or the **Shell** process [70]. The other, known as the **Advancement** or **Fusion process** uses bisphenol A to bring about chain extension in DGEBA [4, 70].

Taffy or Shell process

The Taffy process produces BPA resins of medium to high molecular weight. A low molar ratio (less than three moles [3]) of epichlorohydrin to bisphenol A [24] and a stoichiometric amount of sodium hydroxide [4] are used for the reaction, as shown in Equation 2.40.

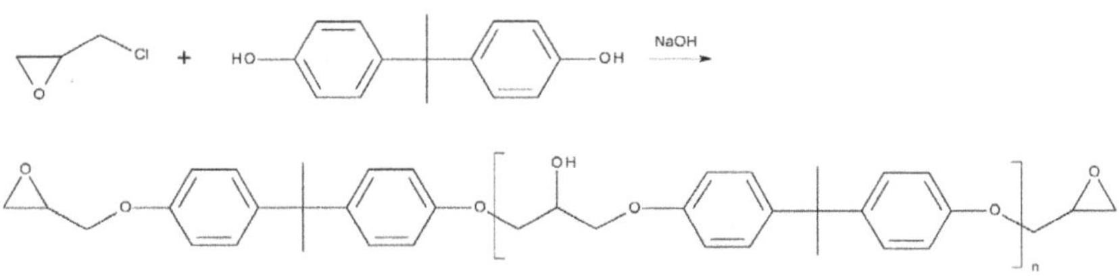

Equation 2.40: Production of epoxy resins by the Taffy process.

The molecular weight of resins yielded by the Taffy process is limited. Table 2.11 shows the stoichiometric ratios and the resultant molecular weights.

Table 2.11: Molecular weights of epoxy resins as a function of molar ratios [4]

Epichlorohdrin: bisphenol A	Repeating units [n]	Molecular weight [g/mol]	EEW * [g]	Melting point [°C]
1.57 : 1	2	900	450–525	65–75
1.22 : 1	3.7	1400	870–1025	95–105
1.15 : 1	8.8	2900	1650–2050	125–135
1.11 : 1	12	3750	2400–4000	145–155

* For a definition of EEW, see Section 2.3

Accordingly, the Taffy process is used in industry up to values of n = 3.7. The reaction yields a two-phase mixture of alkaline sodium chloride solution and aqueous epoxy resin solution. The product is obtained by separating the phases, washing the resin, and removing the water by distillation [4].

Due to the aqueous alkaline milieu, the quantity of a-glycols is relatively high at 0.5 kg per kg resin. These compounds diminish the functionality of the resin and also act as catalyst for amine hardening (see Section 2.4.2). Another disadvantage is the resin's chloride content [4].

Resins made by the Taffy process may contain any number of repeating units (n = 0, 1, 2, 3) because the stoichiometry is not specified [4].

Advancement or Fusion process

So-called solid resins in which n > 2 and which have a high molecular weight are produced by the Advancement or Fusion process [24].

In this reaction, DGEBA (see Equation 2.36) and bisphenol A are made to react at 150 to 190 °C in the presence of a catalyst.

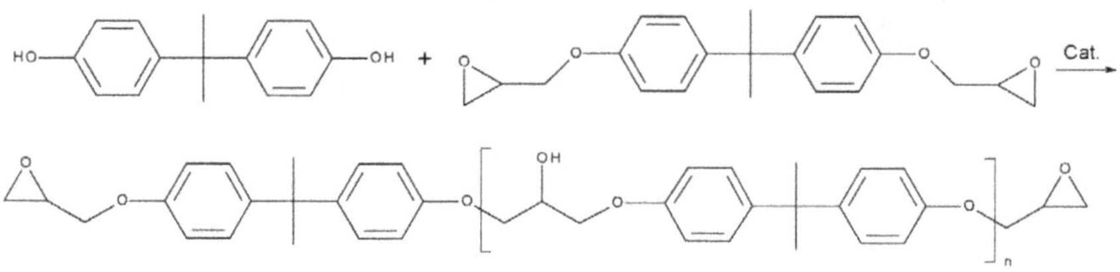

Equation 2.41: Production of epoxy resins by the Advancement Process

To obtain a specific epoxy equivalent weight (EEW; see Section 2.3 for details), i.e. a specified number of epoxy groups per mole of resin, the following equation can be used [24].

$$BisA = \frac{EEW_i^{-1} - EEW_f^{-1}}{EEW_i^{-1} + PEM^{-1}}$$

Formula 2.1: How to calculate the quantity of BPA needed to obtain a specific number of epoxy groups per molecule

Here, BisA is the mass fraction of BPA before the reaction, EEWi is the EEW of the reactant resin, EEWf is the EEW resulting after the reaction and PEM is the BPA equivalent mass and is equal to 114.1 g. The calculation can also be based on epoxy values (see Section 2.3 on how to convert the different values), i.e. equivalents per 1000 g, for which purpose Formula 2.2 can be used [4].

$$BisA_{1000} = \frac{1000\left(E_i - E_f\right)}{8.77 + E_f}$$

Formula 2.2: Calculation of the quantity of BPA needed to generate a specific epoxy resin

Here, $BisA_{1000}$ is the mass of BPA that must be added per 1000 g DGEBA to obtain a specific epoxy value E_f (equivalent/1000 g), expressed in terms of the epoxy value E_i of the DGEBA used.

The Advancement process differs from the Taffy process in that it only yields resins with an even number of monomers (n = 0, 2, 4) [4, 24].

A range of compounds can serve as catalyst, including NaOH, KOH, Na_2CO_3, LiOH, amines, quaternary ammonium salts, borates, and perchlorates [59]. Of more importance is the selectivity of the catalyst, i.e. the extent to which the phenol group, as opposed to the nascent secondary alcohol group, can be made to preferentially react with the epoxy group.

Side reactions with secondary alcohol groups lead to chain branching instead of chain extension, as shown in Equation 2.42 [4].

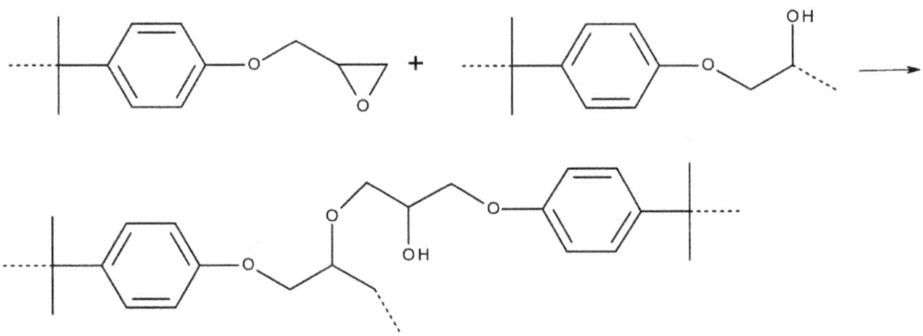

Equation 2.42: Formation of by-products in the reaction between epoxy groups and secondary OH groups

High selectivity is possible with β-hydroxylamines, such as triethanolamine and imidazoles.

The use of numerous aryl or alkyl phosphonium compounds is widely mentioned in patent literature as catalysts for producing very high molecular weights [60].

The high selectivity of phosphonium compounds and the end products of the catalysis can be explained by the following mechanism (Equation 2.43).

Once all the phenolic groups have reacted, the corresponding olefin and trialkyl phosphonium oxide are produced.

The mechanism, which proceeds via the formation of a betaine as intermediate and then of an oxophosphetane which reacts to yield a double bond and phosphine oxide, is familiar from the *Wittig* reaction for the formation of double bonds [68].

The solubility of BPA-based epoxy resins decreases with increase in molecular weight; good solvents for these resins include methyl ethyl ketone (MEK), diacetone alcohol (4-hydroxy-4-methylpentan-2-one), methyl cyclohexanone, methoxy hexanone, ethyl glycol, butyl glycol and ethyl glycol acetate [72, 73].

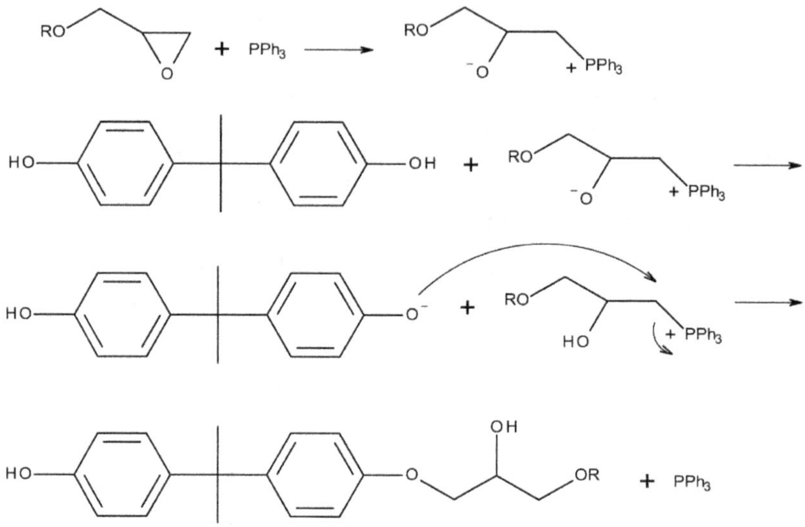

Equation 2.43: Proposed mechanism for the catalysis of triphenyl phosphine in the epoxy resin synthesis [61]

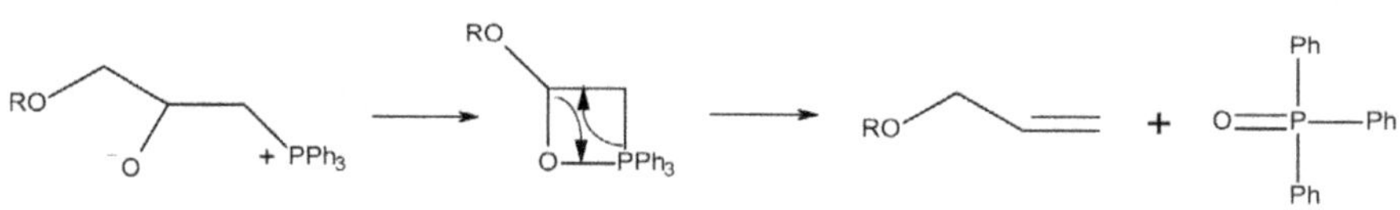

Equation 2.44: Termination reaction for catalysis with triphenyl phosphine [61]

Some viscosity values for epoxy resin solutions in different solvents are presented in Table 2.12.

As mentioned earlier, infrared spectroscopy in the wave number range 400 to 4000 cm^{-1} (mid-infrared or MIR range) affords one way to spectroscopically characterise epoxy resins. Another, increasingly popular method is that of NIR spectroscopy in the wave number range 4000 to 10000 cm^{-1} (near-infrared). In this frequency range, only overtones and combination vibrations of hydrogen bonds (C-H, O-H and N-H) are visible, a fact which simplifies analysis of the spectra [74]. A summary of major resonances of aromatic epoxides is shown in Table 2.13.

In MIR spectroscopy, epoxy resins only have a band at approx. 1250 cm^{-1} (generally 1200 to 1275 cm^{-1}) which is due to the aryl alkyl ether group that is formed [62]. Aliphatic epoxy resins contain a dialkyl ether group that produces bands in the range 1070 to 1150 cm^{-1} [62], and so affords a way of roughly distinguishing between the various epoxy resins.

Table 2.12: Viscosity in mPa s of various epoxy resin solutions

Solvent	Middle molecular weight*			High molecular weight**			Very high molecular weight***
	50 mass.%	40 mass.%	Limit of miscibility	50 mass.%	40 mass.%	Limit of miscibility	50 mass.%
Acetone	< 50	< 50	18	200	50	29	500–700
Methylethylketone	< 50	< 50	∞	250	50	∞	600–900
Methylisobutylketone	50	50	23	1000	-	45	2700–4500
Methylcyclohexanone	250	50	∞	6000	1500	∞	40500
Methoxyhexanone	180	50	∞	3480	580	∞	13500
Diacetonalkohol	250	100	∞	9850–14800	2500	∞	26000
Ethylacetate	< 50	< 50	18	1750	100	28	2000
Butylacetate	50	< 50	20	1300	200	37	4500
Methylglycolacetate	70	< 50	∞	1500	280	∞	7800
Ethylglycolacetate	140	50	∞	1750	500	∞	11000
Isopropylglycolacetate	100	< 50	∞	2500	500	∞	18500
Methylglycol	100	< 50	∞	2000	400	∞	9000
Ethylglycol	140	50	∞	1750–2300	700	∞	13400
Isopropylglycol	125	50	∞	4000	700	∞	22200
Butylglycol	175	50	∞	4600	1000	∞	22100
Methoxyhexanol	350	90	∞	11700	1450	∞	36500
Ethyldiglycol	420	130	∞	2270	1900	∞	37400
Butyldiglycol	550	120	∞	12000	2220	∞	48200

* Values of "Epikote" 1001, ** values of "Epikote" 1007 and *** values of "Epikote" 1009 according to [72, 73]

Table 2.13: NIR signals for epoxy resins [74]

Wavenumber [cm^{-1}]	Wavelenght [nm]	Structural unit
4535	2205	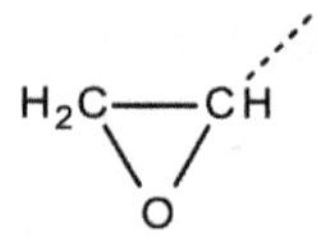
4618	2165	Aromatic group band
4677	2138	Aromatic group band
4784	2090	-OH band
4878	2050	-OH band
5880–5670	1700–1770	-CH overtone of CH_2, CH_3
5988	1670	-CH band of aromatic groups
6060	1650	-CH overtone of 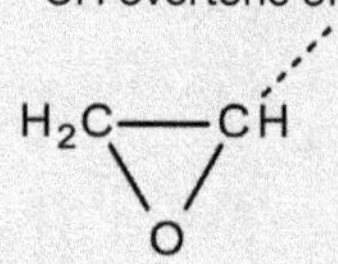
6980	1432	-OH overtone

As mentioned above, NMR spectroscopy reveals structural information about molecules. Table 2.14 shows the ^{1}H-NMR data for DGEBA and Table 2.15 presents the ^{13}C-NMR data for BPA and BPF epoxy resins.

Most publications now use ^{13}C-NMR spectroscopy for characterising epoxy resins, including unknown resins (see [76]).

2.2.2.2 Epoxy novolaks

Both phenol-based (R = H) and cresol-based (R = CH_3) epoxidised novolaks are used in the coatings industry. These two types of resin are produced by acid-catalysed condensation of the phenol with formaldehyde, followed by reaction of the resultant phenol resins with epichlorohydrin [4].

An excess of epichlorohydrin during the synthesis reduces the extent of the reaction between the phenol groups and the epoxy groups and prevents resin branching or crosslinking [4].

Increasing the molecular weight increases the functionality, whereas the opposite holds for BPA-based resins.

Due to the high functionality, the resultant resins have outstanding chemical and thermal resistance and are normally cured at elevated temperatures [4, 7, 24].

Table 2.14: ^{1}H-NMR data for DGEBA

Compound/structural unit	^{1}H-NMR [ppm]	^{2}J [Hz]	cis ^{3}J [Hz]	trans ^{3}J [Hz]
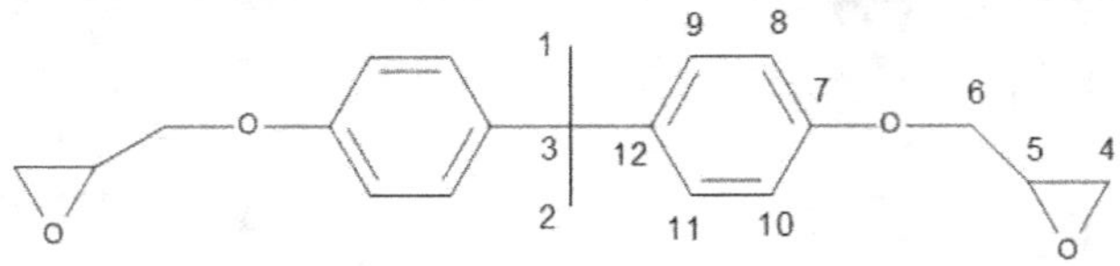	**1.2**: 1.63; **4**: 2.69; 2.82; **5**: 3.28; **6**: 4.27; 3.87; **8.10**: 7.16; **9.11**: 6.86 [69] **1.2**: 1.58; **4**: 2.62; 2.74; **5**: 3.24; **6**: 4.19; 3.84; **8.10**: 7.10; **9.11**: 6.81 [75]	**4**: 5.15; **6**: 11.5 [69]	**4**: 4.25; **5**: 6.35; **8**: 9.56 [69]	**4**: 2.6; **5**: 2.85 [69]

Table 2.15: ^{13}C-NMR data for BPA structural parts

Compound/structural unit	^{13}C-NMR [ppm]
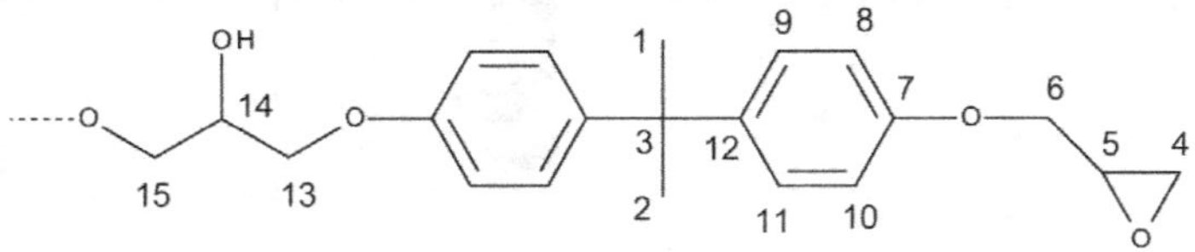	**1.2**: 31; **3**: 62; **12**: 144; **9.11**: 128; **8.10**: 114; **7**: 158; **6**: 68; **5**: 50; **4**: 44; **13**: 70; **14**: 66 [26]
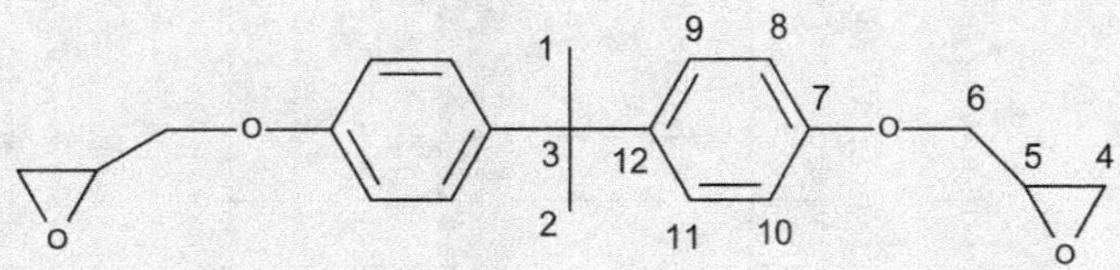	**1.2**: 30.6; **3**: 41.2; **12**: 143.1; **9.11**: 127.3; **8.10**: 113.6; **7**: 155.9; **6**: 68.4; **5**: 49.6; **4**: 44.0 [76]
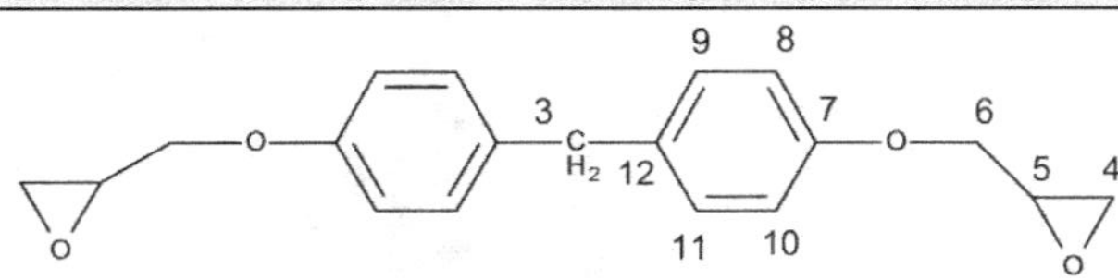	**3**:39.6; **12**: 133.7; **9.11**: 126.9; **8.10**: 114.1; **7**: 156.4; **6**: 68.4; **5**: 49.7; **4**: 44.0 [76]

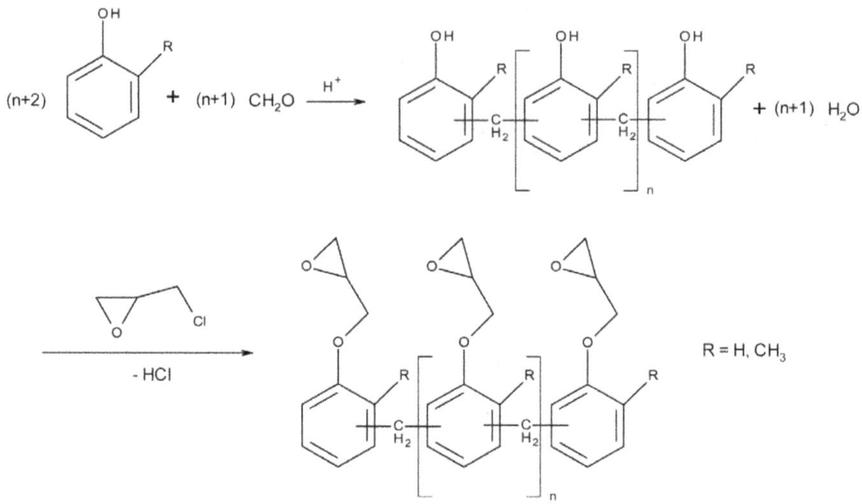

Equation 2.45: Production of phenol and cresol novolaks

Table 2.16: Properties of phenol novolaks

Property	Phenol epoxide novolaks			
	Low molecular weight	Middle molecular weight	Middle molecular weight	High molecular weight
EEW * [g]	176–195 [24]; 175 [4]	178 [4]	200 [4]	185 [4]
Viscosity of the molten mass [Pa*s]	1.5 at 52 °C [24]; 1.4 at 52 °C [4]	20–50 at 52 °C [24]; 35 at 52 °C [4]	3 at 100 °C [4]	0.47–1.35 at 150 °C [24]; 0.8 at 150 °C [4]
Melting point [°C]			53 [4]	73 [4], 66–80 [24]

* EEW is defined in Section 2.3

Table 2.17: Properties of cresol novolaks [4]

Property	Cresol epoxide novolaks			
	Low molecular weight	Middle molecular weight	Middle molecular weight	High molecular weight
EEW * [g]	200	225	229	235
Melting point [°C]	35	73	80	99

* EEW is defined in Section 2.3

2.3 Key parameters of epoxy resins

Several key parameters are used to characterise epoxy resins, most of which do not differ from those used to characterise other resin types. An overview of the most common parameters of resins and methods for determining them is given in DIN 16945:1989. The key parameters of epoxy resins are presented below.

2.3.1 Epoxy equivalent of epoxy resins

The most important parameter for epoxy resins is their content of epoxy groups, because they react with the crosslinking agent and so they determine the amount of crosslinking agent that needs to be used.

There are several methods to determine this content. Table 2.18 provides definitions of widely encountered epoxy parameters. The most common is the epoxide equivalent weight.

Table 2.18: Definitions of various epoxide parameters, according to [7]

Abbreviation	Term	Definition	Sources
EEW	EP-Equivalent	g resin, contains 1 mol epoxy group	DIN EN ISO 3001:1999; DIN EN ISO 7142:2007
EZ	EP-Number	g epoxy groups in 100 g resin	[7]
EW	EP-Value	Mol epoxy groups in 100 g resin	[7, 77]
EI	EP-Index	Mol epoxy groups in 1 kg resin	DIN EN ISO 3001:1999
ES	EP-Oxygen	g epoxy oxygen in 100 g resin	[7, 77]

Table 2.19 shows how the various parameters can be inter-converted.

Table 2.19: Conversion table for various epoxide group parameters, according to [7]

Present/searched	EP equivalent (EEW)	EP number (EZ)	EP value (EW)	EP oxygen (ES)
EP equivalent		$\frac{43 \cdot 100}{EZ}$	$\frac{100}{EW}$	$\frac{16 \cdot 100}{ES}$
EP number	$\frac{43 \cdot 100}{EEW}$		43 EW	$\frac{43 \cdot 100}{ES}$
EP value	$\frac{100}{EW}$	$\frac{EZ}{43}$		$\frac{ES}{16}$
EP oxygen	$\frac{16 \cdot 100}{EEW}$	$\frac{16 \cdot EZ}{43}$	$16 \cdot EW$	

There are several ways to determine the epoxide equivalent weight. The most important methods are presented below. All involve reaction of hydrogen halides with the epoxide group, in accordance with Equation 2.18.

Determination of EEW using HBr

Direct titration of the epoxy resin against HBr in acetic acid in the presence of crystal violet indicator [4] yields good results for glycidyl ethers and cycloaliphatic epoxides. It is unsuitable for glycidyl amines.

Determination of EEW using HCl

The EEW can be determined by reaction of the epoxy groups and HCl in dimethyl formamide solution, followed by titration of the excess HCl against methanolic NaOH solution in the presence of bromophenol blue [77].

DIN 16945 describes a method in which the epoxy resin is dissolved in methyl ethyl ketone and reacts with HCl dissolved in methyl ethyl ketone. The excess HCl is titrated against KOH solution in the presence of cresol red. According to DIN 16945, the EEW is calculated with the aid of the following equation:

$$EEW = \frac{M_E\ 1000}{(a-b)\left[1-\frac{t_1-t_2}{1000}\right]n}$$

Formula 2.3: Calculation of EEW according to DIN 16945

Where: M_E: Initial weight in [g]
a: Consumption of HClO4 (0.1 N) in [ml]
b: Consumption of HClO4 (0.1N) in the blank test in [ml]
t_1: Temperature of the HClO4 during the titration in [°C]
t_2: Temperature of the HClO4 during the adjustment in [°C]

The temperature correction can be omitted if the titration is conducted at constant temperature. Otherwise, the coefficient of expansion of $HClO_4$ must be taken into account.

Determination of EEW using tetraethyl ammonium bromide and perchloric acid

This determination is performed in accordance with DIN 16945 and ISO 3001:1999 by reaction of epoxide groups with tetraethyl ammonium bromide in chloroform/glacial acetic acid solution, followed by titration against perchloric acid in the presence of crystal violet (see also [4]).

DIN 16945 recommends that the more reactive tetrabutyl ammonium iodide serve as the reagent in the case of low-reactivity epoxy resins.

Determination of epoxides that contain amine groups is a two-step process, as described in ISO 3001:1999.

Initially, the amount of amine groups is determined by direct titration against perchloric acid. Then, tetraethyl ammonium bromide and some more perchloric acid are added. The resultant HBr reacts with the epoxy groups and the determination proceeds by titration against perchloric acid in the presence of crystal violet. The EEW is calculated as shown in Formula 2.3.

Determination of EEW using tetrabutyl ammonium iodide and perchloric acid

Initially, the amount of amine groups is determined by direct titration against perchloric acid. Then, tetraethyl ammonium iodide and some more perchloric acid are added. The resultant HI reacts with the epoxy groups and the determination proceeds by titration against perchloric acid in the presence of crystal violet [4].

Determination of EEW using ^{1}H-NMR spectroscopy

Another way to determine the EEW of BPA-based epoxy resins exploits the fact that the signals generated by the aromatic groups and the epoxide group are well separated in the ^{1}H-NMR spectra and therefore can be used to calculate the EEW [83].

The molecular weight M of a BPA-based epoxy resin is related to the repeating unit n as follows.

$$M = 284 \cdot n + 340$$

Formula 2.4: Calculation of the molecular weight of a BPA-based resin with the repeating unit, according to [83]

Here, 284 is the molecular weight of a repeating unit and 340 is the molecular weight of the oligomer in which n = 0.

The EEW can now be calculated according to Formula 2.5.

$$EEW = 142 \cdot n + 170$$

Formula 2.5: Calculation of the EEW based on the repeating units, according to [83]

The repeating unit can be calculated from the intensities of the ^{1}H-NMR signals.

$$n = \frac{\frac{I_2}{I_1} - \frac{I_{2,t}}{I_{1,t}}}{\frac{I_{2,t}}{I_{1,t}}} = \frac{\frac{I_2}{I_1} - 1.33}{1.33}$$

Formula 2.6: Calculation of the EEW based on the repeating units, according to [83]

Where: I_1: Integral of the signals for the epoxy group (2.5 to 3.5 ppm, i.e. three protons)

I_2: Integral of the signals for the aromatic groups (6.8 and 7.1 ppm, i.e. eight protons)

$I_{1,t}$: Integral of the signal for the epoxy group in the oligomer n = 0 (2.5 to 3.5 ppm, i.e. three protons)

$I_{2,t}$: Integral of the signal for the aromatic groups in the oligomer n = 0 (6.8 and 7.1 ppm, i.e. eight protons)

The quotient of the integrals in the oligomer n = 0 has a value of 1.33 and therefore the EEW can be calculated by means of an intramolecular ratio, i.e. the ratio of protons in the aromatic groups to those in the epoxy groups. In the past, internal standards were used, such as 1,1,2,2-tetrachloroethane [84] but this approach require very careful preparation.

2.3.2 Hydroxyl value of epoxy resins

The content of hydroxyl groups in a compound is determined by calculating the hydroxyl value (HV) or hydroxyl equivalent (HE) as described in ISO 7142:2007.

$$HE = \frac{1000}{\frac{5.569\, m_2 + (V_0 - V_1)}{m_1} - \frac{2000}{EEW}}$$

Formula 2.7: Calculation of the OH equivalent in accordance with ISO 7142:2007

Where: HE: g resin per hydroxyl group

m_1: Mass of the sample in [g]

m_2: Mass of pyridinium perchlorate in [g] (see ISO 7142:2007 for details)

V_0: Volume in [ml] of methanolic KOH solution 1M required for the blank test

V_1: Volume in [ml] of methanolic KOH solution 1M required for the determination

EEW: Epoxy equivalent weight in accordance with ISO 3001:1999

$$HV = \frac{56.109 \cdot 1000}{HE} = \frac{56109}{HE}$$

Formula 2.8: Calculation of the OH value in accordance with ISO 7142:2007

The OH value is determined by esterification with acetic anhydride in accordance with Equation 2.46.

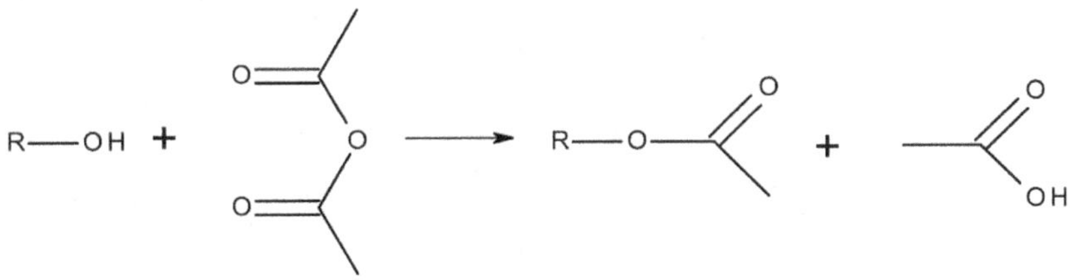

Equation 2.46: The reaction for determining the OH value

The content of free acetic acid correlates with the amount of hydroxyl groups. However, in the case of epoxy resins, the acetic acid reacts with the epoxide groups in accordance with Equation 2.19 to also yield an ester; consequently the EEW (see Section 2.3.1) of the resin must be determined before the OH value is calculated. In Equation 2.4, the reaction of the epoxy group is considered for the calculation of the OH value.

Another way to define the OH value of epoxy resins is based on the reaction of LiAlH4 with alcohol groups [79]. The hydrogen produced during the reaction is measured by the *Zerevitinov* method [80]. The epoxy groups are unable to react under these conditions and do not interfere with the measurement. However, this method is highly complex and is only used rarely.

2.3.3 Chloride content of epoxy resins

All epoxy resins based on epichlorohydrin can contain chlorine by way of impurity, either as chloride or bound to carbon. As impurities critically affect the properties of the resultant coating or the application process, they need to be determined accurately. Table 2.20 presents the most common chlorine-containing impurities of epoxy resins, as listed in the three-part ISO 21627:2010 standard.

Determinations of chlorine impurities can distinguish between inorganic chlorine, readily-saponifiable chlorine, and total chlorine.

Table 2.20: Structures of the common chlorine-containing impurities in epoxy resins, according to ISO 21627

Structural unit	Name	Group	DIN
Cl^-	Chloride	Inorganic chlorine	DIN EN ISO 21627-1:2010
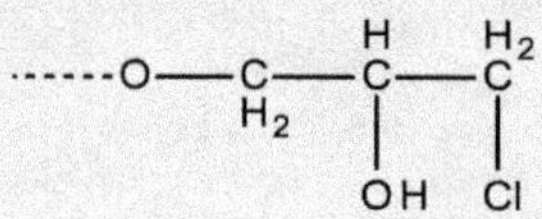	1,2-Chlorohydrin	Easy saponified chlorine	DIN EN ISO 21627-2:2010
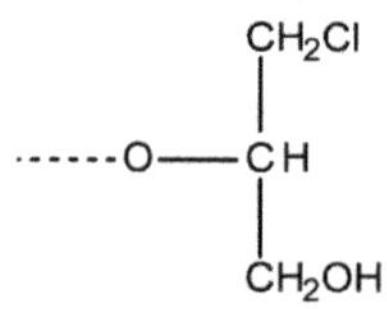	1,3-Chlorohydrin	Easy saponified chlorine	DIN EN ISO 21627-2:2010
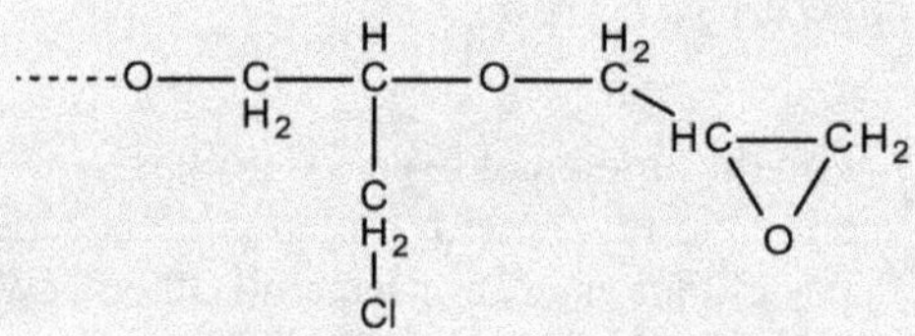	1-Chlor-2-glycidyl ether	Easy saponified chlorine	DIN EN ISO 21627-2:2010

Inorganic chlorine

The resin is dissolved as described in ISO 21627-1:2010 in a suitable solvent, such as acetone, methyl ethyl ketone or tetrahydrofuran, and titrated potentiometrically against silver nitrate to determine the chlorine content.

Easily saponifiable chlorine

As described in ISO 21627-2:2010, the epoxy resins are made to react with butoxy ethanol in NaOH solution, with release of bound chlorine. This solution is acidified with glacial acetic acid and titrated potentiometrically against silver nitrate solution to determine the chlorine content; the inorganic chlorine content must be subtracted from the result.

Total chlorine content

The resin is dissolved in diethylene glycol monobutyl ether and heated under reflux in alcoholic NaOH solution so that the resin is completely saponified (see ISO 21627-3:2010).

The resultant solution is titrated potentiometrically against silver nitrate solution to determine the chlorine content.

2.3.4 Determining the tendency of liquid resins to crystallise

Liquid epoxy resins, especially BPA-based types, have a more or less pronounced tendency to crystallise. By comparison, isomeric mixtures of BPF-based resins generally undergo no crystallisation.

This tendency can only be determined qualitatively, and one possibility is proposed in ISO 4895:1999, in which the resin is mixed with calcium carbonate and some ethanol and stored for several hours at 10 °C. The tendency to crystallise is assessed visually at regular intervals. For the purposes of comparison, the resins are classified as (a) liquid, (b) has crystals but is flowable, and (c) is fully crystallised and solid.

2.3.5 Detection reactions

Shell describes two colour reactions for distinguishing between those resins which contain BPA and those which do not.

Test 1

1 ml of conc. HNO_3 (ca. 63 %) is added to 1 ml of a solution of the resin in sulphuric acid. The mixture is shaken and allowed to stand for five minutes. The solution is then stirred into 100 ml of a 5 % NaOH solution. The presence of an epoxy resin based on diphenylol propane or its ester is indicated by the development of an orange-red colour [78].

Test 2

To prepare Denigés reagent, 10 ml conc. sulphuric acid is dissolved in 50 ml of water. Then, 2.5 g of mercury oxide is dissolved in the hot solution. If the solution is not clear, it must be filtered.

Then 5 ml of Denigés is reagent added to 1 ml of the resin solution in sulphuric acid. The solution is shaken thoroughly and allowed to stand for 30 minutes. The presence of an epoxy resin based on a diphenylol propane or a derivative thereof is indicated by the development of an orange-red precipitate [78].

These colour reactions need the presence of the hydroxyl groups in the BPA-based resins, as BPA on its own, like aliphatic and cycloaliphatic epoxy resins, does not exhibit a reaction. Only BPA- (and also BPF)-based epoxy resins containing some hydroxyl groups react positively.

2.4 Structure and properties of polymers based on epoxy resins and their curing processes

2.4.1 Polyether polyols and phenoxy resins

Phenoxy resins, as mentioned in Section 1.3.1, are polyether polyols which are usually produced by the Taffy or Advancement process (Section 2.2.2) and which do not contain any detectable epoxy groups [7]. Nevertheless, resins that contain the OH group are also considered epoxies; for the purpose of this book, though, they are referred to as phenoxy resins. These polymers may contain as many as 20 to 30 repeating units and have molecular weights of up to 45,000 g/mol – this stands in contrast to resins which contain epoxy groups and whose molecular weights do not exceed 8,000 g/mol [4, 7]. As shown in Table 2.10, these resins fall into the very high molecular weight category.

As phenoxy resins have no epoxy groups, the resins are thermally stable and serve as thermoplastics [4]. In the coatings industry, the resins are cured with crosslinking agents, which avail of the numerous OH groups in the molecules. Some key curing reactions are presented in the following.

Curing with phenol-formaldehyde resins

One common method of curing phenoxy resins utilises etherified phenol-formaldehyde resins, which are produced by reaction of phenols and formaldehyde in the ratio of 1 : 2 to 1 : 3, and subsequent reaction between several active methylol groups and aliphatic alcohols [81]. Etherification has three effects:

1. The stability of the crosslinking agent increases as a result of the lower reactivity
2. The solubility or compatibility with the resins increases
3. The alcohols increase the elasticity of the resultant coating [82]

The resins can be hardened in 1-pack systems under acid catalysis, forming a network with phenoxy resins at temperatures above 150 °C [71]. This is the preferred combination for can coatings (see Section 3.5).

Curing with melamines

In a manner reminiscent of the reaction between phenoxy resins and phenol formaldehyde resins, phenoxy resins can be made to react with etherified melamine-formaldehyde resins. The resultant networks find application in numerous coating systems.

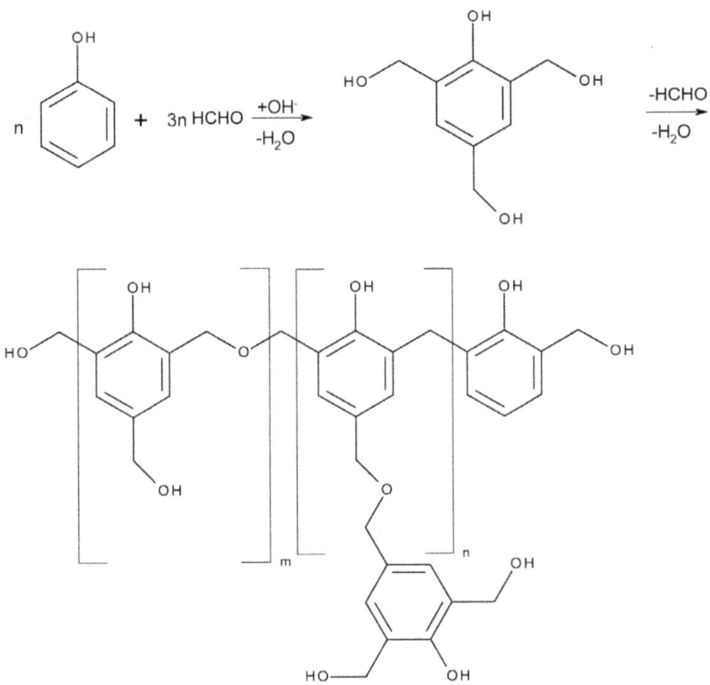

Equation 2.47: Production of phenol-formaldehyde resins with idealised structures, according to [82]

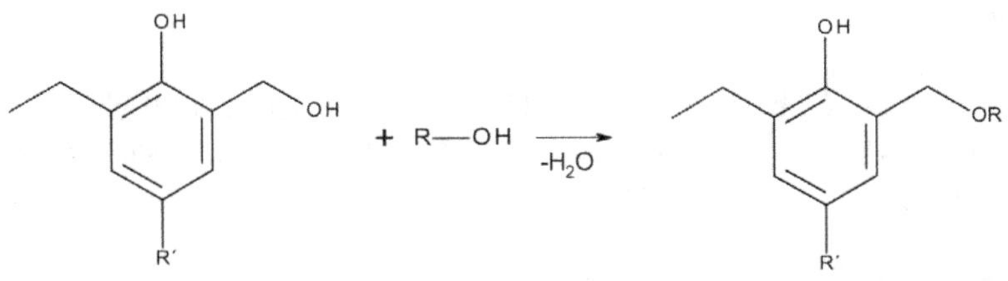

Equation 2.48: Etherification of phenol-formaldehyde resins with aliphatic alcohols, according to [82]

The 1-pack systems can be crosslinked at 130 to 150 °C under acid catalysis and produce light-stable coatings that are very tough [71].

Curing with isocyanates

Another way to crosslink phenoxy resins is to make them react with isocyanate groups to yield polyurethane networks (PUR).

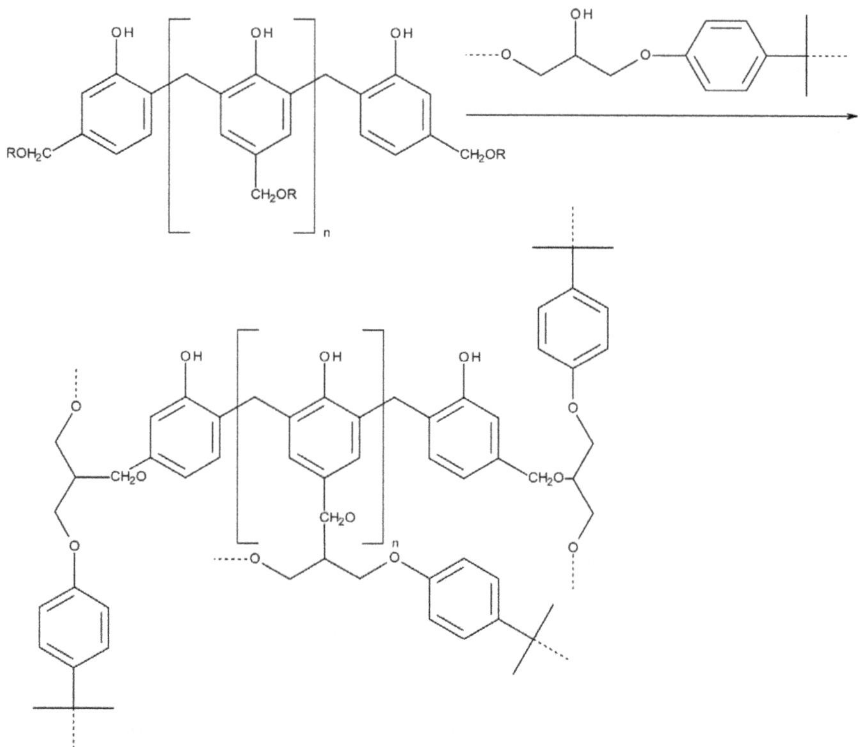

Equation 2.49: Reaction between phenoxy resins and phenol-formaldehyde resins, according to [4]

The 2-pack systems can be crosslinked at 15 to 50 °C to yield coatings of very high toughness, adhesion and mechanical strength. Such coatings are therefore used in metal coatings for machinery [71].

2.4.2 Polyether polyols with epoxy groups

Epoxy resins with molecular weights of up to 8,000 g/mol actually contain epoxy groups and are therefore "genuine" epoxy resins, which as mentioned earlier may be produced by the *Taffy* or *advancement* process. A detailed description of these resins is given in Table 2.10.

The resins contain two types of functional groups, namely hydroxyl and epoxy, and both are used to form networks in coatings. The reaction and underlying mechanism are presented in Section 2.1. In the following, we will take a closer look at some uses in crosslinking reactions.

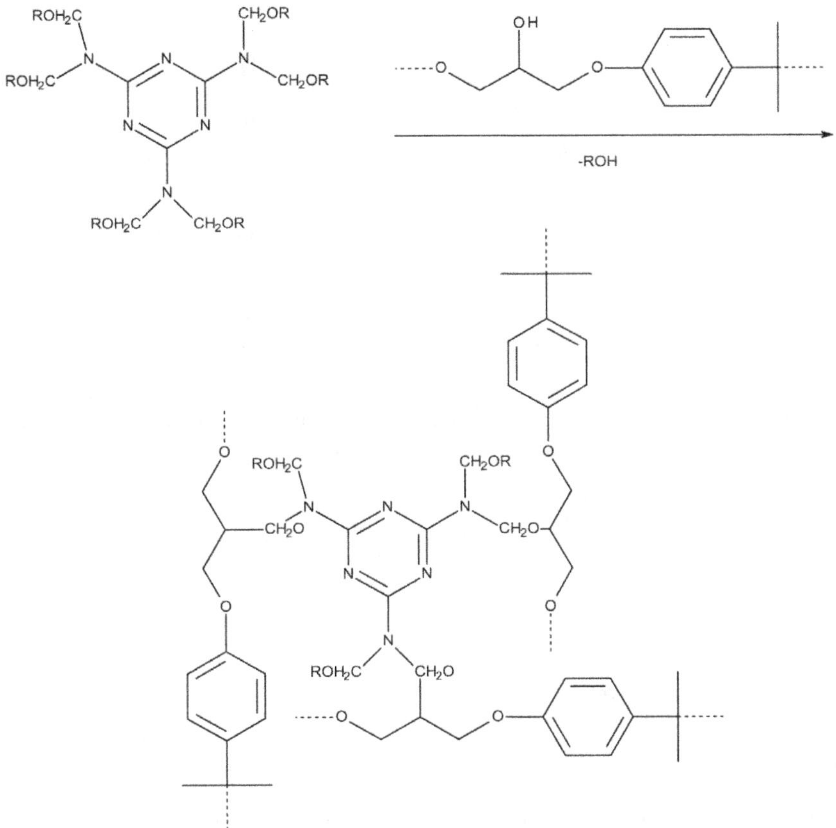

Equation 2.50: Reaction between phenoxy resins and melamine-formaldehyde resins, according to [4]

2.4.2.1 Catalytic curing of epoxy resins

By catalytic curing is meant homopolymerisation of the epoxy resin to a branched polyether polyol with the aid of either *Lewis* acids or *Lewis* bases.

Lewis bases

Lewis bases for homopolymerisation of epoxy resins are normally tertiary amines, such as 2,4,6-tris-(N,N-dimethylaminomethyl)phenol and N,N-dimethylbenzylamine [3].

The zwitterion formed in Equation 2.6 is not stable under the curing conditions and reacts with the OH group to yield an 1,2-hydroxylamine and an alcoholate anion.

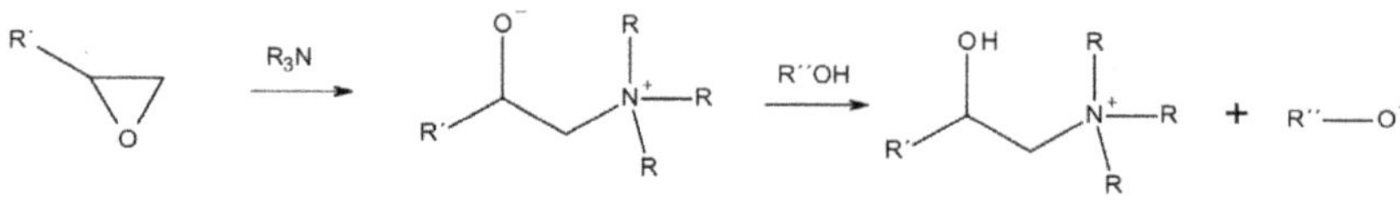

Equation 2.51: Reaction between phenoxy resins and isocyanates to yield polyurethanes

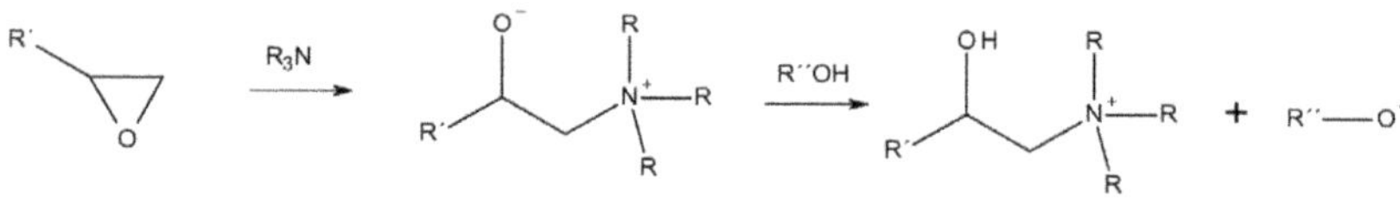

Equation 2.52: Initiation of the homopolymerisation by tertiary amines, according to [4, 7]

The alcoholate anion is sufficiently basic, especially under aprotic conditions, to react with another epoxy ring and so initiate the polymerisation.

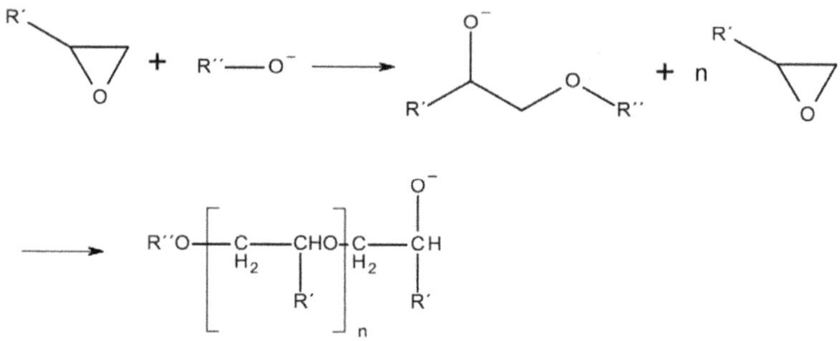

Equation 2.53: Chain propagation of the homopolymerisation by amine catalysis [4, 7]

The effectiveness of the homopolymerisation depends not only on the temperature but also on the concentration and structure of the tertiary amine, especially the extent of steric hindrance at the nitrogen atom.

Lewis acids

The Lewis acid most commonly employed for homopolymerisation is boron trifluoride (BF_3), which crosslinks to yield a network in a matter of a few minutes.

Complexes of BF_3 with amines prolong the pot-life, yielding systems that cure at 80 to 100 °C. A common Lewis acid catalyst is BF_3-methylethylamine complex, which decomposes above 80 to 100 °C and releases the catalyst [3]. The mechanism behind the homopolymerisation of the complexes presumably involves an oxonium ion which is formed by epoxide-induced solvation of the boron trifluoride-amine complex and which initiates chain growth due to polarization by another epoxide group.

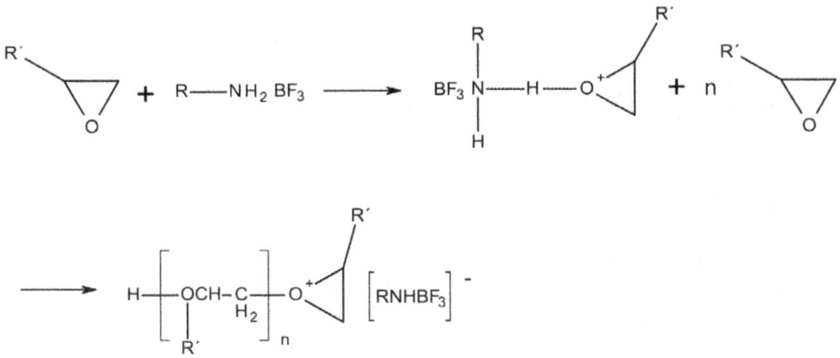

Equation 2.54: Initiation of homopolymerisation by boron trifluoride complexes, according to [4]

Besides *Lewis* acids, superacids such as trifluoromethanesulphonic acid (F_3CSO_3H), hexafluoroantimonic acid ($HSbF_6$) and hexafluorophosphoric acid (HPF_6) can serve as catalysts for the homopolymerisation [67].

As very strong acids contain very weak conjugate bases, the latter are not nucleophilic.

Only superacids make effective catalysts for the homopolymerisation because the conjugate bases of strong acids such as HCl or H_2SO_4 are nucleophilic – they preferentially enter into a substitution reaction to yield the chlorohydrin or sulphonate and so inhibit homopolymerisation.

Homopolymerisation can also be performed at elevated temperatures by blocking the superacids with amines. One example is α,α-dimethylbenzylpyridinium hexafluoroantimonate, which can be cured at 120 °C with aliphatic epoxy resins [67, 88].

The possibility of producing or releasing latent acids photochemically is covered in Section 3.1.2.

Curing with amines

Section 2.1.1 explains the reactions between epoxy groups and amines (Formula 2.4) and the catalytic effect of hydroxyl groups (Formula 2.5) on the reaction between amines and epoxides. From the equation for the reaction, it follows that the number of possible reactions with epoxy groups is equal to the number of hydrogen atoms on the nitrogen. The functionality of amines can therefore be described by the H-equivalent mass, as shown in Formula 2.9.

Where: M: Molecular weight of the amine
H: Number of active hydrogen atoms

$$H - Equivalent = \frac{M}{H}$$

Formula 2.9: Calculating the H-equivalent of amines, according to [4]

Another way of characterising amines utilises the amine number as defined in DIN 16945:1989.

The amine number defines the amount of KOH in milligrams that is equivalent to one gram of the substance, as determined by titration against perchloric acid according to the following formula.

$$Amine\ number = \frac{a \cdot F \cdot 5.6}{m}$$

Formula 2.10: Calculating the amine number according to DIN 16945:1989

Where: a: Consumption of perchloric acid, $c(HClO_4) = 0.1$ mol/l
F: Factor (titre) of the perchloric acid, $c(HClO_4) = 0.1$ mol/l
M: Initial weight of amine in g

The quantity of amine groups or the H-equivalent can also be determined according to DIN EN ISO 9702:1998 with a distinction being possible between primary, secondary and tertiary amines. The ability to distinguish between them is important because the different amines produce different network points.

A primary amine can react with two epoxide groups to produce a node whereas a secondary amine can only lead to chain extension, and a tertiary amine serves merely as a catalyst and does not produce a network.

Total amine content

The total amine content of aliphatic and aromatic amine hardeners is the sum of primary, secondary and tertiary amines, as determined by potentiometric titration against HBr or $HClO_4$, with perchloric acid being unsuitable for amine hardeners, such as N-aminoethyl piperazine.

Content of primary amines

The content of primary amines in aliphatic curing agents is determined by converting the amines to imines with a defined excess of 2,4-pentanedione (acetyl acetone) in DMF; the excess 2,4-pentanedione is determined by titration against KOH; imines have a neutral reaction under these conditions.

The number of amine groups in aromatic curing agents is calculated by subtracting the content of secondary and tertiary amines from the total amine content.

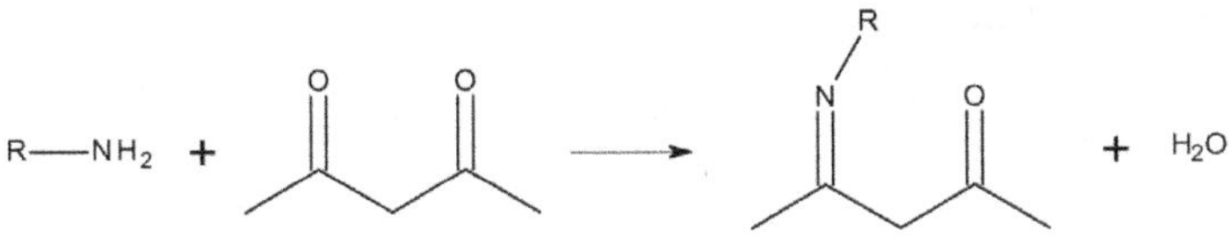

Equation 2.55: Reaction for determining primary amine groups in aliphatic crosslinking agents, according to DIN EN ISO 9702:1998

Content of secondary amines

The content of secondary and tertiary amine groups in aromatic curing agents is determined after reaction of the primary amines with salicylaldehyde in glacial acetic acid to yield a Schiff's base.

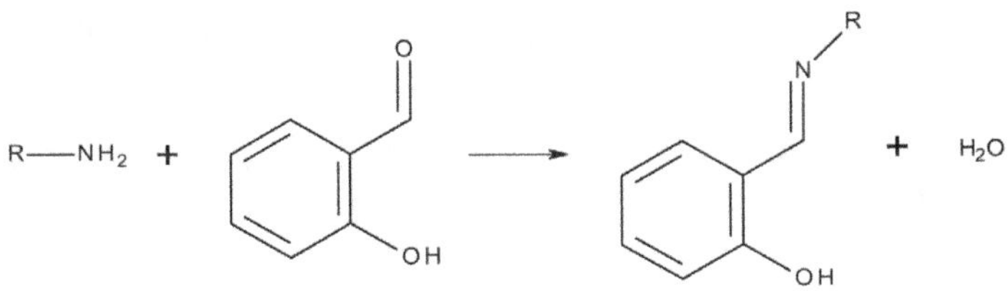

Equation 2.56: Reaction for determining secondary and tertiary amine groups in aromatic hardeners, according to DIN EN ISO 9702:1998

Strongly basic amines are then determined by potentiometric titration against HCl in glacial acetic acid while HBr or $HClO_4$ is used in the case of weakly basic amines, such as 4,4-diaminodiphenylsulphone (see Table 2.21).

The content of secondary amine groups in aliphatic curing agents is calculated by subtracting the content of primary and tertiary amine groups from the total content.

Content of tertiary amines

The content of tertiary amines for both aromatic and aliphatic curing agents is determined by potentiometric titration against HBr or $HClO_4$, with primary and secondary amine groups first being converted with acetic anhydride to amides, which do not subsequently interfere with the analysis.

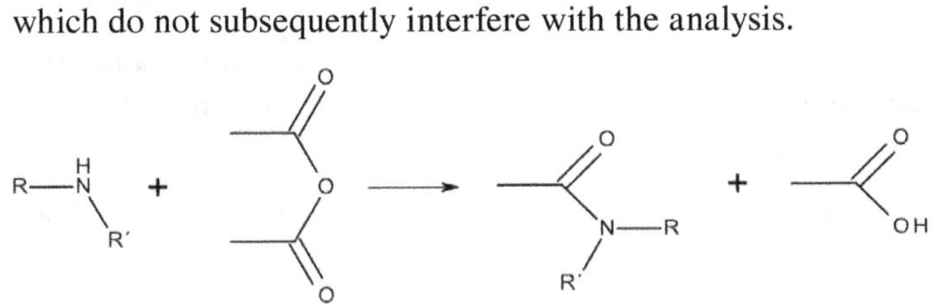

Equation 2.57: Reaction for determining tertiary amine groups in amine-hardeners, according to DIN EN ISO 9702:1998; R′= H or alkyl/aryl

Amine hardeners and their properties

Normally, epoxy resins are cured with aliphatic polyamines, but some aromatic amines are used, too.

An overview of the amines commonly employed is presented in Table 2.21.

Aromatic amines are much less basic than aliphatic amines, and so need elevated temperatures in order to react with epoxy resins. In contrast, aliphatic amines such as ethylenediamine, diethylenetriamine etc. react at ambient conditions without an accelerator and have pot-lives of 15 to 30 minutes [4, 14].

Cycloaliphatic and aromatic amines need curing temperatures of 120 to 140 °C [14]. Aromatic amines react faster than aliphatic amines with cycloaliphatic epoxides, however, because they are more acidic. The aliphatic amines cannot cure at ambient conditions and need elevated temperatures for curing to occur.

In general, sterically bulky groups on the nitrogen diminish the reactivity. So also do electron-withdrawing groups, which diminish the nucleophilic properties of the amine nitrogen [4].

Network formation during the reaction between amines and epoxy resins is critically dependent on the stoichiometry of the two components and the catalysis. The glass transition temperature (T_g) of the resultant network will reach a maximum if the reactants are used in a ratio of 1 : 1. If one component is present in excess, the T_g is much lower [91, 92].

Adding a catalyst such as imidazole [91] or tertiary amines [93] accelerates the reaction between amine and epoxy and also initiates or accelerates the reaction between the hydroxyl groups and the epoxy (see Section 2.1.1). The reaction between hydroxyl groups and epoxy groups leads to a significant reduction in the T_g.

The disadvantages of aliphatic and aromatic polyamines are a pronounced sensitising effect on human skin, difficult-to-achieve mixing ratios of 10 : 100 and the short pot-life (often less than two hours) of mixtures of aliphatic amines and epoxy resins.

Furthermore, amine hardeners generally lead to poor surface hardness and slow, full cure. The surfaces remain tacky for a long time and cannot be mechanically loaded.

Finally, the surfaces are prone to spotting due to formation of carbamate with atmospheric CO_2 [4, 7].

One way to avoid these problems is to use polyamine-epoxy adducts, i.e. products made by the reaction of excess polyamine with an epoxy resin.

Table 2.21: Amine hardeners for epoxy resins

Name	Abbre-viation	Structure	Viscosity at 25 °C [mPas]	Mo-lecular weight [g/mol]	H-active equivalent mass [g]	Properties
Ethylendiamine	EDA	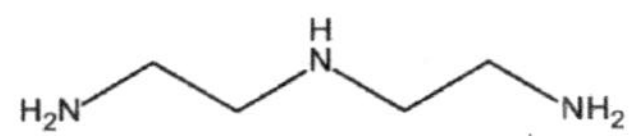	1.3–1.5	60.1	15	Because of the toxicologic properties (acut. Tox. 3) the use is questionable
Diethylentriamine	DETA	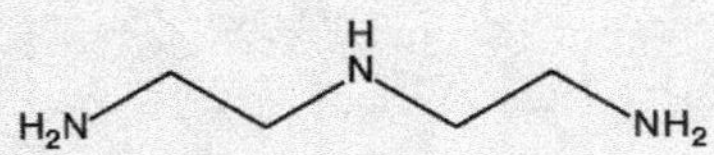	5–10	103.1	21	Tends to yellowing and produce under cold curing tacky and susceptible to moisture surfaces [7]
Triethylentriamine	TETA	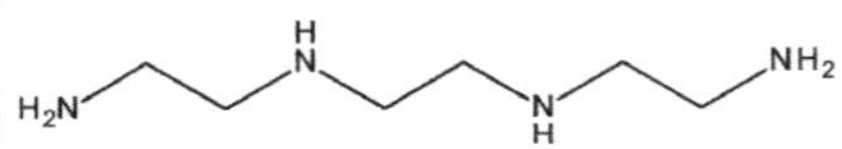	20–30	146.1	25	Tends to yellowing and produce under cold curing tacky and susceptible to moisture surfaces [7]
3,3′,5-Trimethylhexa–methylendiamine	TMD	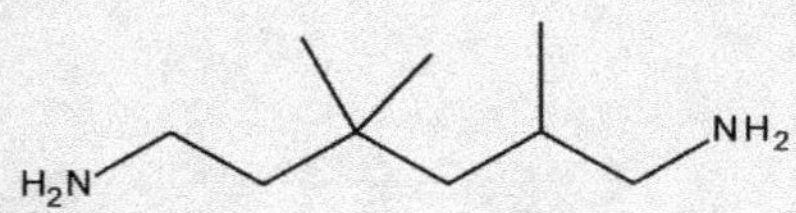	3–5	143.1	39.6	Outstanding light resistance and high reactivity [7]

Table 2.21 Continue

Name	Abbre-viation	Structure	Viscosity at 25 °C [mPas]	Mo-lecular weight [g/mol]	H-active equivalent mass [g]	Properties
1,2-Cyclohexyldiamine, 1,2-Diaminocyclohexane [4]	DAC [4]	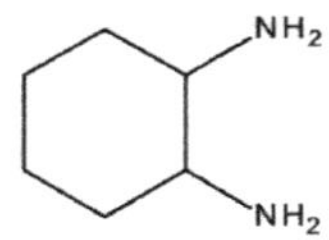	10–14	114.1	28	Tends to yellowing and produce under susceptible to moisture surfaces with good mechanical and chemical resistance [4,7]
Isophorondiamine	IPD	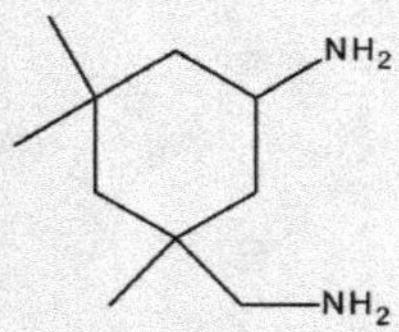	10–15	170.2	42.5	Good light resistance and produce under susceptible to moisture surfaces with good mechanical and chemical resistance [4,7]
4,4´-Diamino-3,3´-dimethyldicyclohexylmethane		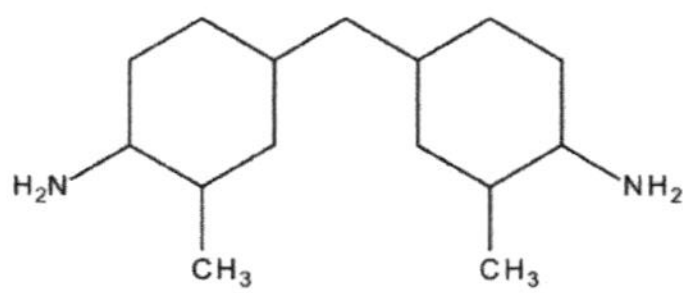	80–100	238.2	60	High light resistance but low reactivity causes curing at elevated temperature [4,7]
N-Aminoethylpiperazine	AEP	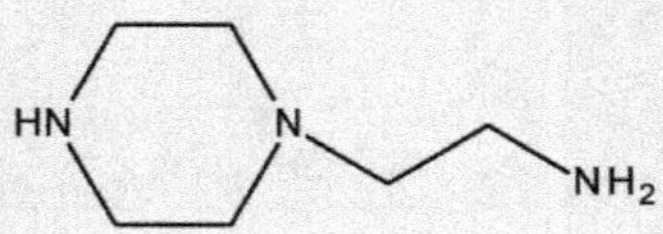	15 [86]	129.1	43	

Table 2.21 Continue

Name	Abbreviation	Structure	Viscosity at 25 °C [mPas]	Molecular weight [g/mol]	H-active equivalent mass [g]	Properties
m-Phenylendiamine		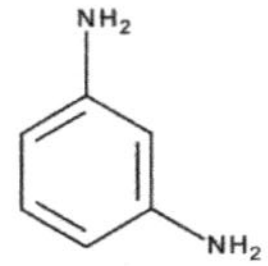	M.P. 138–141 °C [87]	108.1	27	
m-Xylylendiamine	MXDA	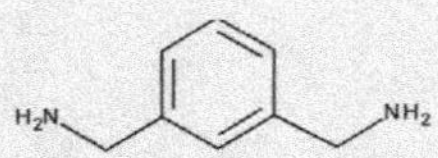	5–8	136.1	34. 32 [85]	[7]
Methylendianiline	MDA	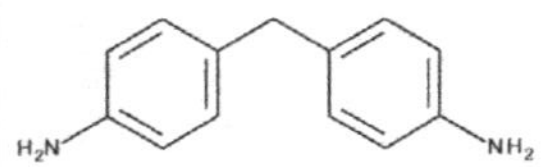	M.P. 90 °C	198.1	50, 50 [85]	High tend to yellowing and toxicological (Carc. Cat. 2, Muta.Cat 3) very questionable [7]. Because of the low reactivity a curing at 120–140 °C is necessary [14]
4,4-Diamino-diphenylsulfone	DDS	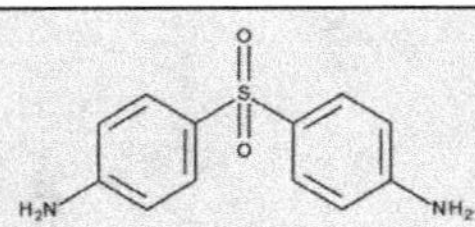	M.P. 172–175 °C	248.1	62, 62 [85]	Because of the low reactivity a curing at 120–140 °C is necessary [14]
Dicyandiamide	DICY	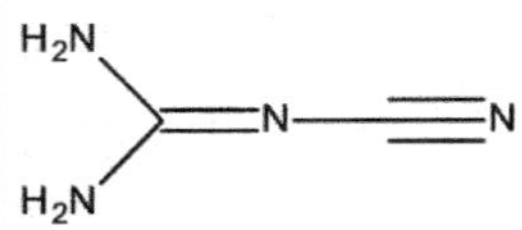	M.P. 205 °C [13]	84	28	Because of the poor solubility in epoxy resins mixtures are storage stable and could use for powder coatings [14] and one pack systems [4], whereas the resulting surfaces shows good electrical properties and a high temperature stability [4]

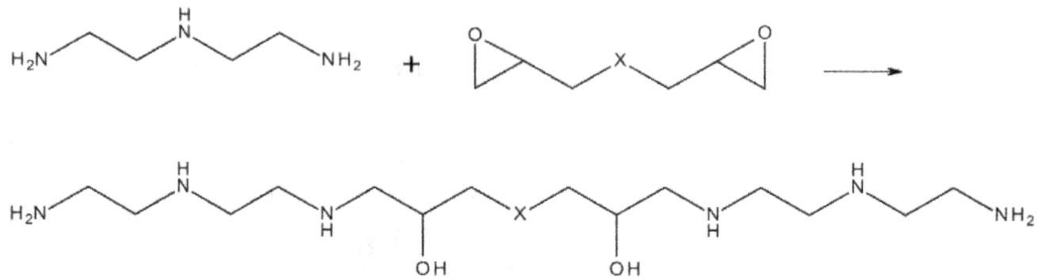

Equation 2.58: Production of epoxy-amine adducts; X = bisphenol, alkyl etc.

This consists in making an excess of aliphatic amine, such as DETA, react with DGEBA and removing the excess amine by distillation. The outcome is an amine with a higher H-equivalent, which both lowers the mixing ratio and mitigates the sensitising effect. Furthermore, the hydrophobisation reduces disadvantageous water absorption during curing and also decreases the extent of carbamate formation [4, 7, 67]. Epoxy resins can be hardened with polyamine adducts at 15 °C and with a catalyst at 5 °C; they have a pot-life of 30 to 60 minutes [71].

Polyamines can react with ethylene oxide to yield polyether polyolamines. For example, DETA reacts in the presence of water with ethylene oxide to afford a mixture of mono- and di-hydroxyethyl diethylenetriamine.

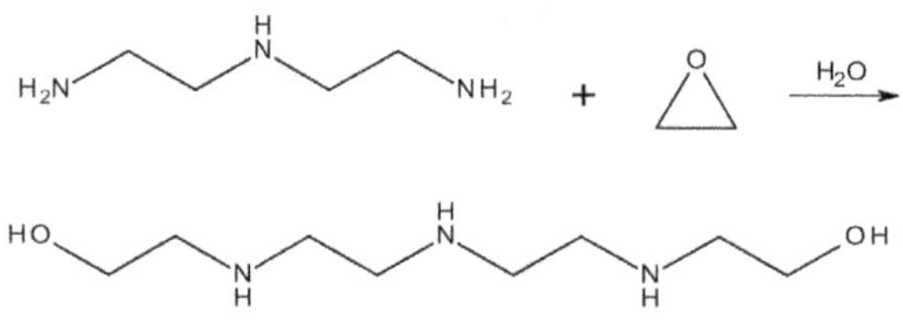

Equation 2.59: Reaction between ethylene oxide and DETA, according to [4]

The pot-life can be extended with ketimines, which are produced by making polyamines react with ketones, usually methyl ethyl ketone (MEK). As ketimines cannot react with epoxy groups until they have been hydrolysed by atmospheric moisture, they may be considered a type of blocked amine [4, 67].

Ketimine- or aldimine-blocked amine hardeners have curing times of less than 16 hours and can be stored for up to 200 days, provided that moisture is excluded [14].

After application, the ketimine reacts with atmospheric moisture to release the amine; release is boosted by the low boiling point of MEK and the attendant rapid shift in equilibrium. Nonetheless, ketimines are not used in 1-pack systems but rather only in 2-pack systems that have a long shelf life. The reason for this is the hard-to-remove residual water content in the solvents and pigments [67].

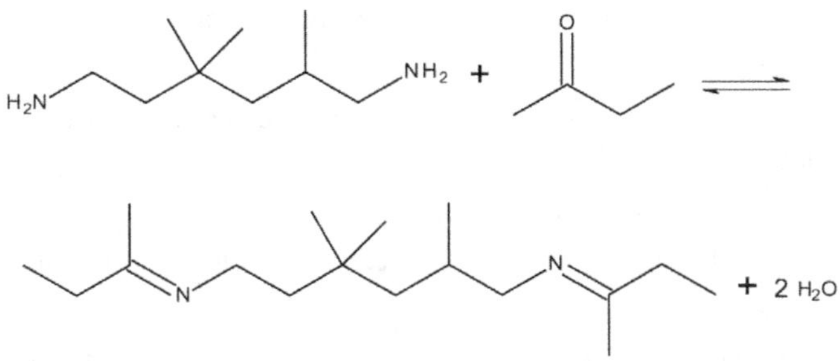

Equation 2.60: The equilibrium of ketimine formation, as illustrated by the reaction between MEK and TMD

The most common amine hardeners are polyamides produced by the reaction between dimerised or trimerised vegetable fatty acids and polyamines.

Dimerised fatty acids are obtained by a *Diels-Alder* reaction [95] involving 9,12- and 9,11-linoleic acids.

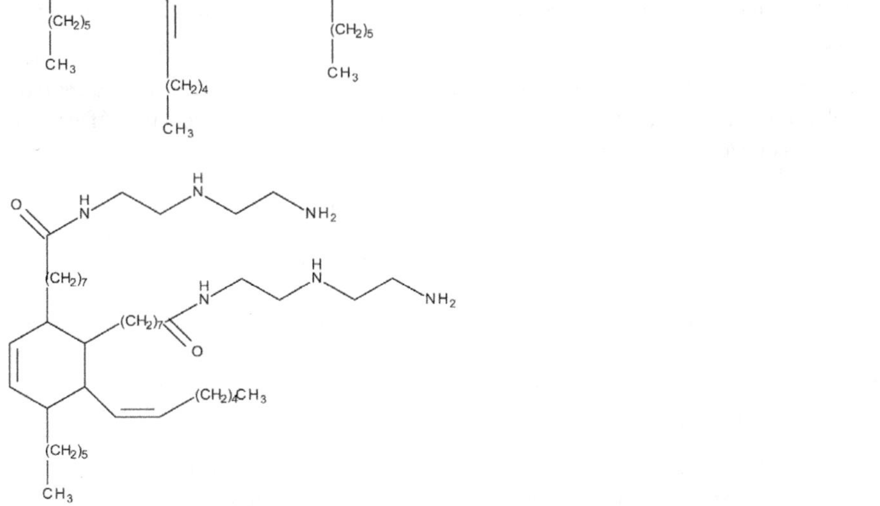

Equation 2.61: Production of polyamides from dimerised fatty acids and polyamines [4, 96]

If the dimerised fatty acids are made to react with polyamines, complex mixtures of polyamides are formed.

DETA, TETA, aminoethylpiperazine and aromatic diamines such as m-phenylenediamine serve as the amines.

As the amide group cannot react with the epoxy group, only the (usually) terminal amine groups are available for reaction with the resin. This lowers the reactivity significantly.

The polyamides primarly used in coating materials are notable for their good compatibility, low reactivity and consequently long pot-lives, and mild curing conditions (from 15 °C and above). They yield flexible coatings that offer good adhesion, high moisture resistance and a low tendency to form carbamate [4, 7, 14, 67, 71].

If DETA is used for the production of polyamides, terminal imidazoline groups may be formed through cleavage of water, as shown in Equation 2.62.

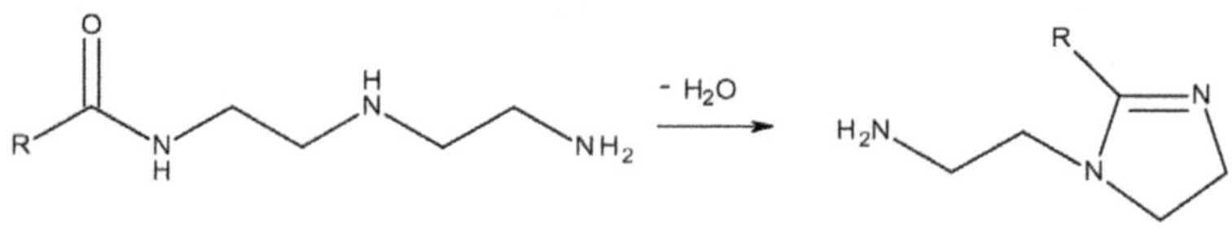

Equation 2.62: Formation of imidazolines from DETA polyamides

These groups in turn lower the reactivity and affect the solubility and compatibility in the formulation. Imidazoline contents of 35 to 85 % are available industrially; this enables the properties of the hardened coating to be finely adjusted [67].

Amidoamines have properties similar to those of polyamides, but have a much lower viscosity. They are produced by reaction of fatty acids or other monocarboxylic acids with polyamines, and subsequent reaction to amides and, after ring closure, to imidazolines (see Equation 2.62) [4, 67].

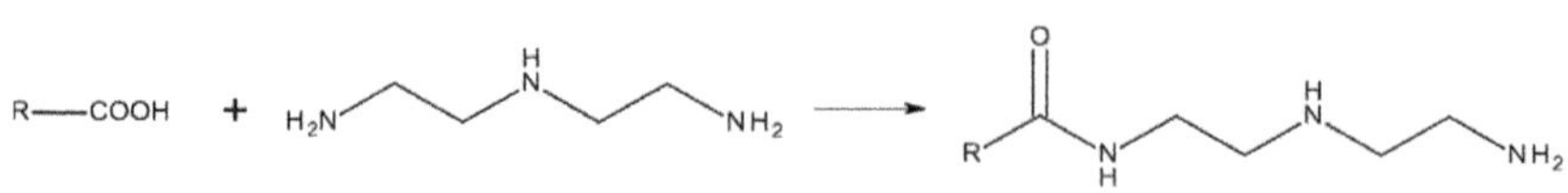

Equation 2.63: Production of amidoamines from fatty acids and DETA

Another group of amine hardeners is the *Mannich* bases produced by reaction between phenols, formaldehyde and a polyamine.

Mannich base hardeners are highly reactive and exhibit good full cure properties from 5 °C and above, yielding coatings with outstanding chemical resistance. Owing to their high reactivity, they have a short pot-life [7].

Table 2.22: Properties of coatings based on DGEBA with amines, according to [4]

Property		Aliphatic polyamine	Aliphatic polyamine adduct	Aromatic polyamine	Aromatic polamine adduct	Polyamido-amine	Polyamide	Cyclo-aliphatic polamine
Gel time, 100 g at 25 °C [min.]		< 30	< 30	> 120	> 120	30–60	>120	30–60
Viscosity at 25 °C [mPa s]	Crosslinking agent	5–50	200–3000	2–5000	200–3000	200–3000	> 8000	< 200
	Resin and crosslinking agent	> 3000	3000–10000	10000–20000	2–6000	3000–6000	> 10000	< 3000
Curing time at 25 °C [h]		10–15	< 10		10–15	< 10	15–24	< 10
Deflection temperature [°C]		80–120		142		60–80	40–60	
Breaking force [MPa]		40–70		83		59–69	35–55	
Exural strength [MPa]		104–124		152		104–117	48–62	
Compressive strength [MPa]		138–227		207		90–104	62–76	
Dieletric constant at 50 Hz and 25 °C		4–5	4–5	4–5	4–5	3.2–3.6	3.2–3.6	
Specific volume resistance at 25 °C [W cm]		1016	1016	1016	1016	1016	1016	1016

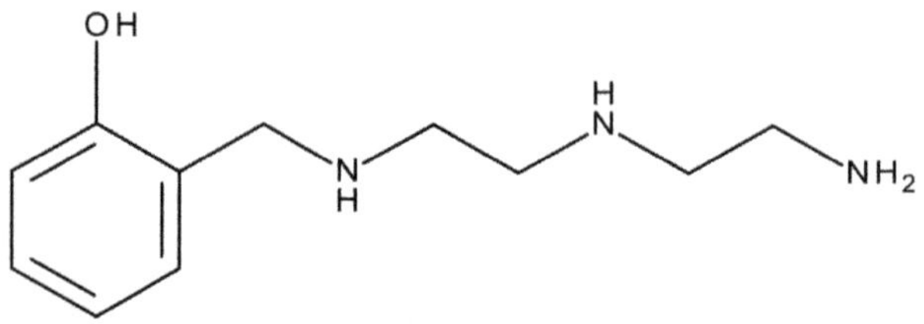

Equation 2.64: DETA-phenol Mannich base

The numerous reactions of dicyandiamide (DICY) and its poor solubility are covered in Section 2.1.1. Thanks to these properties, DICY occupies an exceptional position among amine hardeners [14].

Properties of amine-epoxy networks

Some properties of DGEBA/amine coatings are presented in Table 2.22 to illustrate the effects of various amine hardeners.

Generally, epoxy-amine coatings are highly sensitive to acids, especially acetic acid, but are stable to bases. Acetic acid penetrates into the coating and forms acetates with the amine groups, producing a hydrophilic surface and increasing water uptake, which has a negative impact on corrosion-protection properties. Nevertheless, a 10 % excess of amine is necessary for ensuring that all the epoxy groups react in full [14, 67].

Table 2.23: NIR signals for epoxy resins and reaction products with amines [74]

Wavenumber [cm^{-1}]	Wavelength [nm]	Structural unit
4618	2165	Aromatic group band
4677	2138	Aromatic group band
4784	2090	-OH band
4878	2050	-OH band
4995	2002	-NH band
5045	1982	$-NH_2$ band
5880–5670	1700–1770	-CH overtone of CH_2, CH_3
5988	1670	-CH aromatic group band
6536	1530	$-NH_2$ overtone
6635	1507	$-NH_2$, -NH overtone
6980	1432	-OH overtone

Epoxy-amine networks and epoxy-amine adducts can be readily characterised by spectroscopic methods.

One such method is NIR spectroscopy, already covered under epoxy resins (see Table 2.15). In this wavelength range, only OH and NH groups generate signals. Consequently, tertiary amines are not visible, nor do they interfere with the measurement (Table 2.23). It is worth noting that primary amines can be distinguished from secondary amines, which is important for characterising epoxy-amine adducts.

Curing with alcohols

Provided that a suitable catalyst is used, epoxy resins can serve as hardeners for polyol resins, at temperatures of 120 °C [67]. The reactions involved are covered in Sections 2.1.1 and 2.1.2 (see Equations 2.11 and 2.22).

The compounds can also be analysed by means of ^{1}H- and ^{13}C-NMR, which is especially useful in the case of epoxy-amine adducts (see Tables 2.24 and 2.25)

Table 2.24: ^{1}H-NMR data for epoxy-amine adducts

Compound/structural unit	^{1}H-NMR [ppm]	cis ^{3}J [Hz]
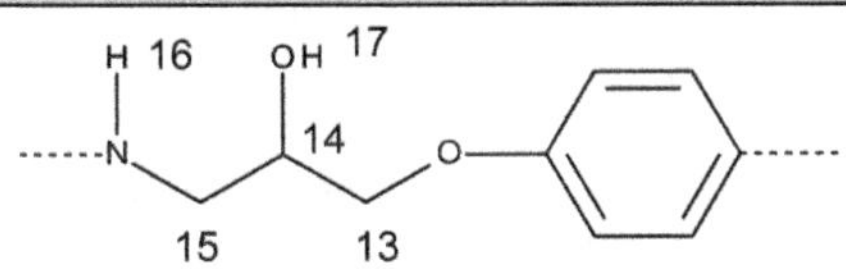	**13**: 3.9; **14**: 4.02; **15**: 3.09; 3.25; **16**: 5.04; **17**: 5.09; 5.12 [75]	**17**: 9.8 [75]

Table 2.25: ^{13}C-NMR data for BPA structural units in epoxy-amine adducts and amine-epoxy resins

Compound/structural unit	^{13}C-NMR [ppm]
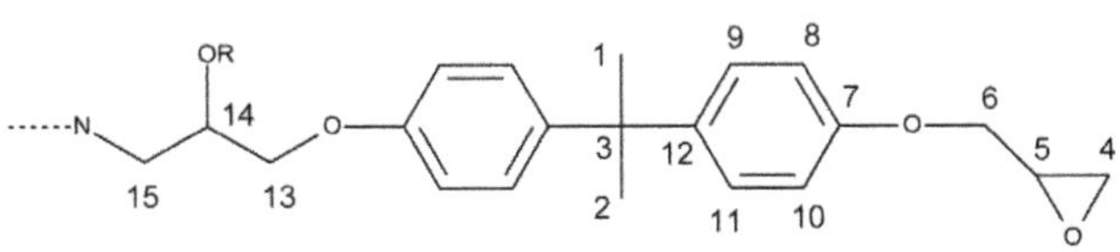	**1.2**: 31.3; **3**: 42.2; **12**: 143.9; **9.11**: 128.1; **8.10**: 114.5; **7**: 157.5; **6**: 69.9; **5**: 50.2; **4**: 44.2; **13**: 71.1; **14**: 68.1; **15**: 56.0/57.7 [29] **1.2**: 31; **3**: 41.7; **12**: 143.5; **9.11**: 127.7; **8.10**: 114; **7**: 69.7; **6**: 68.7; **5**: 50.1; **4**: 44.6 [65] vgl. [89.90]
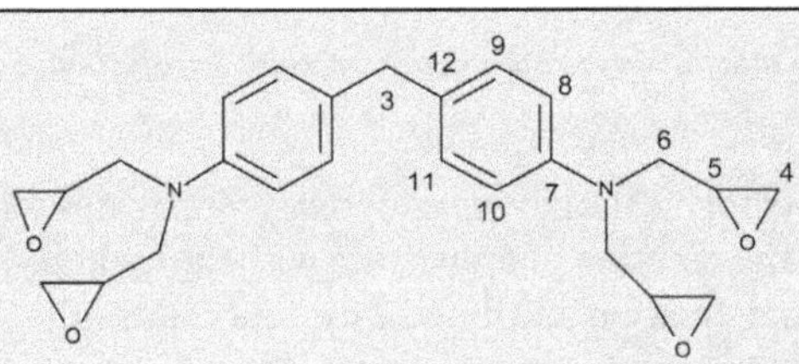	**3**: 39.2; **12**: 130.0; **9.11**: 129.0; **8.10**: 112.1; **7**: 146.1; **6**: 52.5; **5**: 49.9; **4**: 44.7 [76]

Curing with anhydrides

Anhydrides do not possess active hydrogen and so cannot react direct with epoxy groups. The reaction of anhydrides with epoxy resins requires elevated temperatures – in 1-pack systems 150 °C and above [71] – and entails converting the hydroxyl group on the epoxy resin into the hemiester. A free carboxylic group is formed that reacts with an epoxy group to yield an ester, as described in Section 2.1.2 (Equation 2.19), enabling a network to be built up (Equation 2.65). In the absence of hydroxyl groups on the epoxy resin, the reaction can be initiated with a catalytic quantity of butanol because every reaction between a carboxylic group and an epoxy group produces a new hydroxyl group [14].

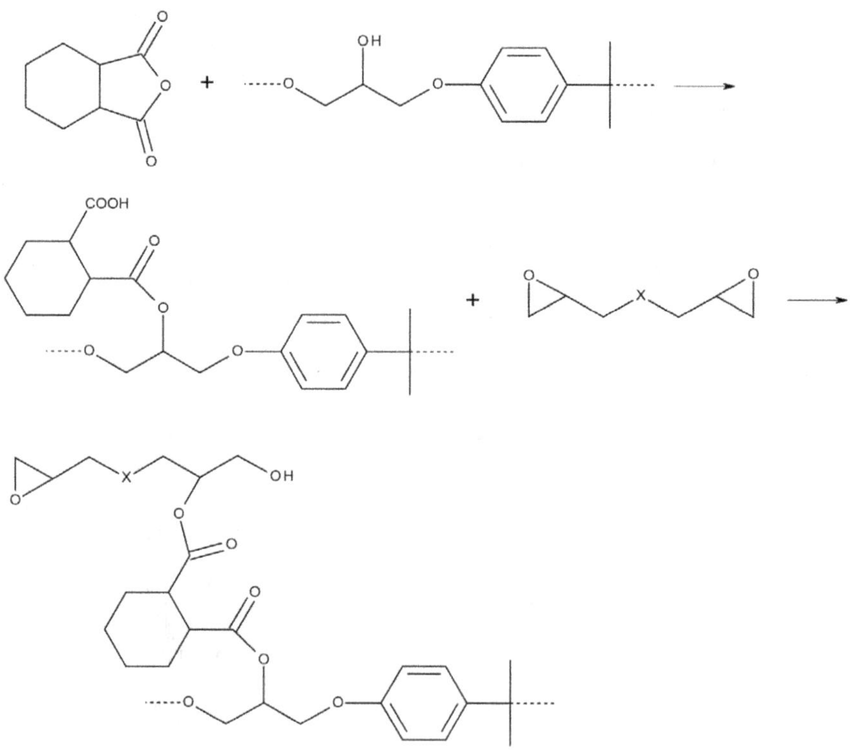

Equation 2.65: Mechanism behind anhydride curing of epoxy resins; X = bisphenol, alkyl etc.

If ammonium salts are used to catalyse the curing of epoxy group resins with anhydrides, they merely accelerate the reaction of anhydride with the hydroxyl group – reaction between carboxylic and epoxy is inhibited by ammonium salts. This catalysis therefore can therefore lower the network density [94].

As is the case for amine hardeners, the activity of anhydride curing agents is determined by the number of anhydride groups. The content of reactive anhydride groups or the total content of acid groups is set out in DIN 16945.

The content of anhydride groups is measured as hemiester formation by titrating against methanolate and phenolphthalein in acetone, and the total content of acid groups is determined by saponifying the anhydride groups with NaOH, followed by titration against HCl and a thymolphthalein indicator.

The results from the titrations can be used to calculate the parameters as shown in Formula 2.11.

$$\text{Content of acid anhydride} = \frac{(b-c)F_2\,M_A}{M_{E2}\,100} - \frac{a\,F_1\,M_A}{M_{E1}\,100}$$

$$\text{Content of acid} = \frac{a\,F_1\,M_S}{M_{E1}\,100} - \frac{(b-c)F_2\,M_S}{M_{E2}\,200}$$

Formula 2.11: Calculation of the anhydride content and acid content, according to DIN 16945.

Where: M_A: Molecular weight of the anhydride
M_S: Molecular weight of the acid
M_{E1}: Initial weight in [g] in the sodium ethylate titration
M_{E2}: Initial weight in [g] in the saponification
a: Consumption of sodium ethylate solution (0.1 M) [ml]
b: Consumption of HCl (0.1 M) [ml] in the blind test
c: Consumption of HCl (0.1 M) [ml] in the test
F_1: Factor for the sodium ethylate solution
F_2: Factor for the HCl

Another important parameter is the content of free acid in the anhydride curing agent, because it defines the reactivity and storage stability, especially in 1-pack systems.

The content of free acid groups in the anhydride is determined in accordance with ISO 7327:1997 by photometric measurement of the curing agent using rhodamine 6G in a toluene/MEK solution.

The resultant coatings are sensitive to saponification because polyesters are formed, but the surfaces are relatively stable to acids. Using aliphatic or cycloaliphatic anhydrides instead of their aromatic counterpart's yields surfaces that possess greater insulating properties, even under long-term heat load [14].

Table 2.26: Anhydride curing (crosslinking) agents

Name	Abbreviation	Structure	H-equivalent weight [g]	Properties
Pyromellitic dianhydride	PMDA	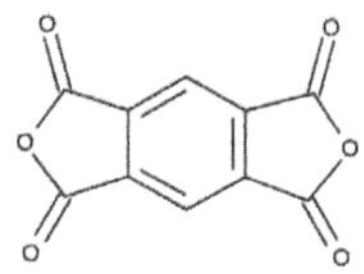	109	Crosslinking agent for powder coatings [85]
Phthalic anhydride	PSA	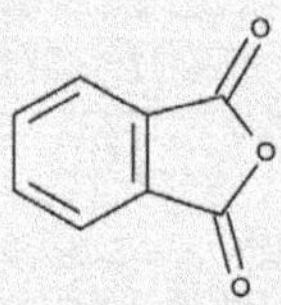	148	
Tetrahydrophthalic anhydride	THPSA	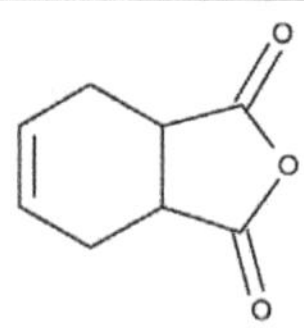	152	Hot curing crosslinking agent for casting resins [85]
Hexahydrophthalsäureanhydrid	HHPSA	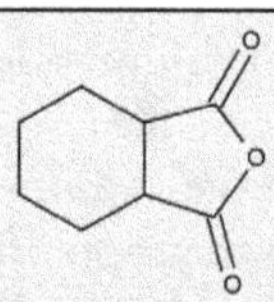	154	Hot curing crosslinking agent for casting and laminating resins [85]
Methyl-hexahydrophthalic anhydride	MHHPSA	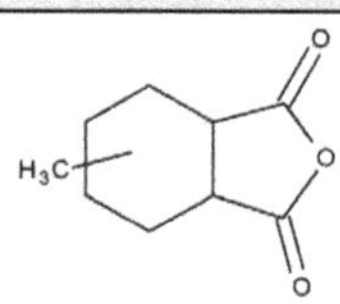	168	Crosslinking agent for epoxide resin adhesives and thermosetting coatings [85]
Admergin acid		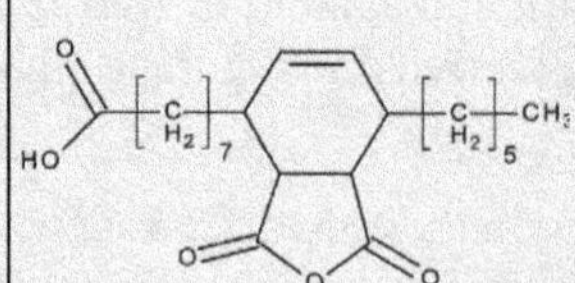	370	Trifunctional carboxylic acid (Diels-Alder product of maleinic anhydride with unsaturated conjugated fatty acids), thermosetting crosslinking agent for liquid epoxide resins [85]

Curing with alcohols

The reaction between epoxy groups and thiols to yield β-hydroxysulphides is covered in Section 2.1.1 (Equation 2.14).

The reaction can be catalysed by amines; the thiolate group serves as the reactive intermediate, combining with epoxy groups in ambient conditions. This type of catalysis is used in 2-pack systems [67].

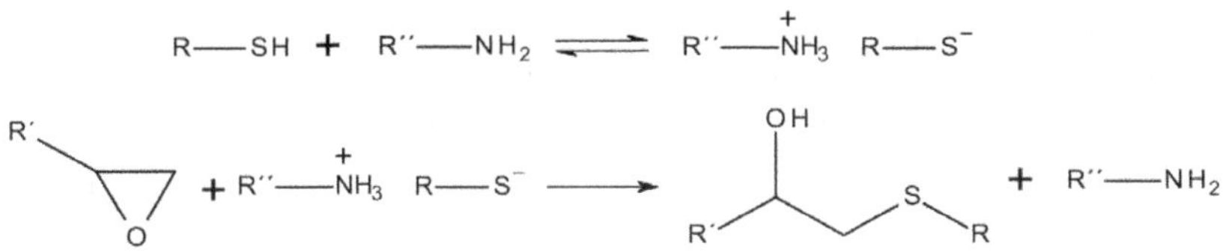

Equation 2.66: Amine catalysis of the reaction between thiols and epoxides

A special group among thiol crosslinking agents is that of polysulphides, which have the general structure shown in Equation 2.67 [98].

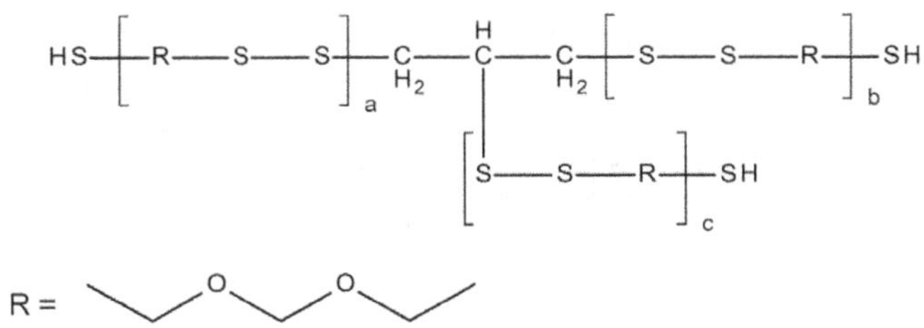

Equation 2.67: Structure of polysulphides, according to [98]

Resins used in the sealants industry are cured by oxidation with manganese(IV) oxide [98], but they can also be cured with epoxy resins in the presence of amine catalysts to yield flexible thioether networks.

What makes these curing agents special is the possibility of removing strain from the network by changing the disulphide bridges. If pressure is exerted on the network (or if strain arises during curing or as a result of thermal treatment), the network breaks by rupturing the disulphide bridges, which can then close again to relieve the pressure or strain in the network.

The choice of curing mechanism depends on the intended use. Although there is no single optimal curing reaction that covers all kinds of applications, there are so many variations that it is always possible to obtain the desired property profile. Some properties of coatings created with different curing agents are presented in Table 2.27.

Table 2.27: Properties of networks produced from epoxy resins (DGEBA [3, 6] A/F: crystallisation-free bisphenol A/F mixture with 20 % hexanediglycidyl ether [7]) and different curing agents

Resin	DGEBA	DGEBA	DGEBA	DGEBA	DGEBA	DGEBA	DGEBA	A/F	A/F	A/F	A/F	A/F
Crosslinking agent	**TETA**	**DETA***	**MDA**	**Poly-amide**	**An-hy-dride****	**HHPSA**	**BF_3-MEA**	**DETA**	**IPD**	**Poly-amino-amide**	**Adduct cross-linking agent**	**Mannich-base cross-linking agent**
Accelerator					Dimethyl-benzylamine							
Mixing ratio – Resin	100	100	100	100	100	100	100	100	100	100	100	100
Mixing ratio – Crosslinking agent	13	25	26	43	87.5	83	3	12	25	55	53	28
Curing temperature [°C]	RT	RT	80 +140	RT	100 (4h)	120–160	100 (4h)	RT	RT	RT	RT	RT
Glass transition temperature [°C]		55	158			136						
Heat deflection temperature [°C]	111		160	101	156		168	54	52	30	35	55
Martens value [°C]		45				111						
Modulus of elastivcity [MPa]	3.05	3.05	3.5 [6]; 2.7 [3]	2.14	3.05	3.5	3.1	3.8	3.9	1	1.6	2.7
Dielectric loss	0.02	0.015	0.004 [6]; 0.015 [3]	0.007	0.0054	0.004	0.0053					
Dieletric constant at 10^3 Hz	3.9		4.06	3.19	3.14		3.45					

DETA*: With DETA modified with dibutyl phthalate; anhydride**: methylbicyclo[2.2.1]heptene-2,3-dicarboxylanhydride; MEA: methylethylamine; polyamide: Versamide 140 (Henkel Corp.)

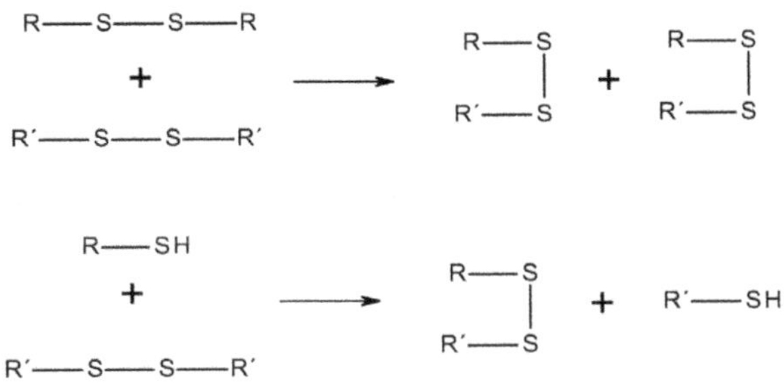

Equation 2.68: Bond swapping by disulphide bridges and between thiols and disulphides [99, 100]

2.4.3 Waterborne epoxy resins

There are basically two ways to make a resin compatible with water. The first uses emulsifiers to stabilise the polymer in water by adsorbing the emulsifier onto the polymer, and the second incorporates ionic or hydrophilic groups into the polymer to yield self-emulsifying resins.

The latter method can use anionic, cationic or non-ionic groups (see Equation 2.69).

Waterborne epoxy resins are prepared by emulsifying conventional epoxy resins with suitable emulsifiers or, more commonly, by dispersing epoxy resins that contain water-soluble groups, such as carboxylate or protonated amines [3]. Both are presented in the following section.

Waterborne epoxy resin systems stabilised with emulsifiers

Waterborne epoxy-amine coating systems can be prepared via emulsions. Emulsifiers can be added to the resin formulation or to both the resin and the amine hardener, to allow water to be used to thin the coating and to lower the solvent content [67]. Waterborne epoxy resin coatings that cure at room temperature can be produced by combining emulsified BPA-based liquid resins with salts of polyamines or polyamides [4]. Furthermore, protonated polyaminoamides can serve as both curing agent and emulsifier for the epoxy resin [7]. Self-emulsifying resins are obtained by making reactive emulsifiers react with the epoxy resin or with the curing agent, which is often a polyamidoamine.

Glycidyl esters of versatic acid (see Section 2.2.1 and Table 2.8) can serve as the reactive emulsifiers. The curing agents can be obtained via reaction of polyether-functionalised carboxylic acids and polyamines or polyamidoamines, with formation of amide (see Equation 2.70). These two strategies can be combined to produce coatings that have the same performance profiles as conventional BPA-based coatings [110].

Anionic polymers:

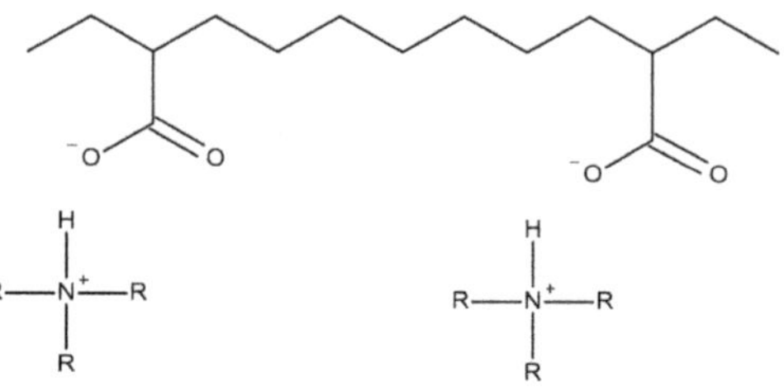

Cationic polymers:

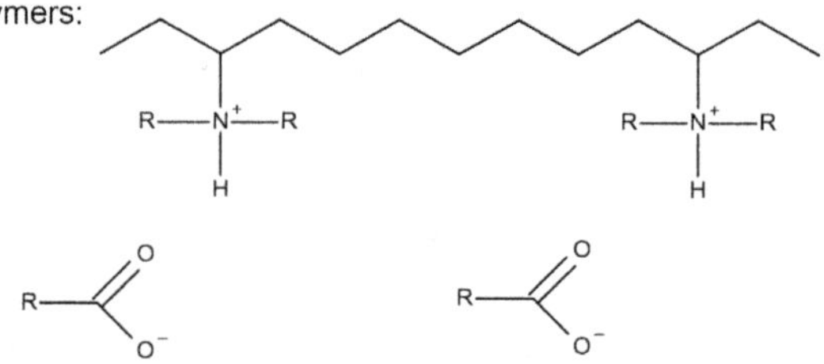

Hydrophilic polymers:

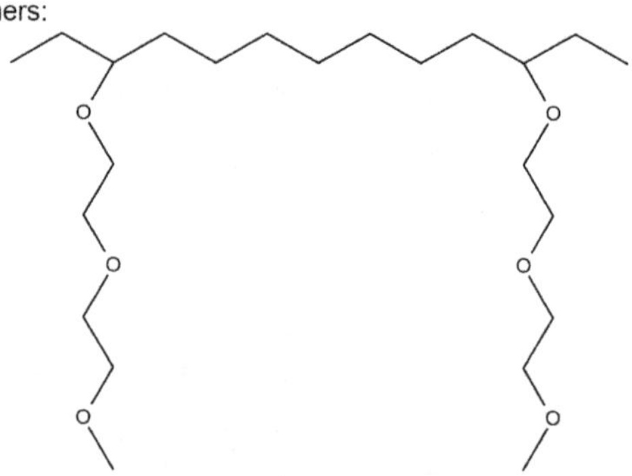

Equation 2.69: Schematic diagram of the various methods for effecting stabilisation by attaching functional groups to the polymer

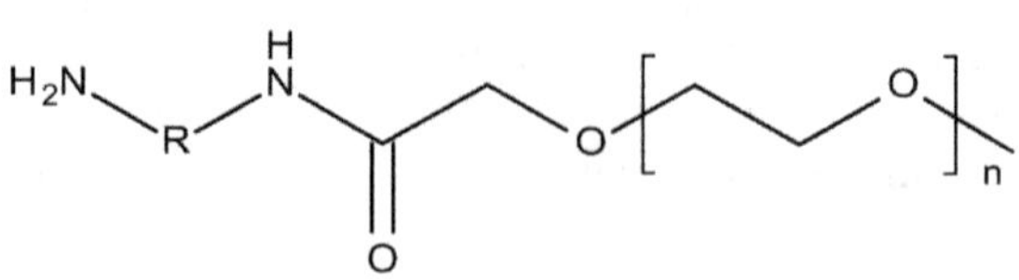

Equation 2.70: Structure of self-emulsifying polyamidoamine curing agents, according to [110]; R = alkyl or aryl

Furthermore, readily sandable, waterborne fillers can be obtained by combining epoxy resins with special amine-urethane crosslinking agents which have been made to react with epoxy-functional emulsifiers [105].

Converting low-molecular epoxy resins with monohydroxy-functional polyethylene or polypropylene oxides or their blends in the melt to the corresponding ethers yields solvent-free emulsifiers. These can be combined with epoxy resins to obtain emulsions, without the need for co-solvents [106].

A variant of this synthesis is to add amines to produce emulsifiable crosslinking agents that exhibit very little shrinkage as they cure [109].

Waterborne epoxy resin systems with ionically stabilised dispersions

An obvious way to introduce ionic groups into epoxy resins is the production of epoxy-amine adducts followed by neutralisation of the resin with carboxylic acids. The resultant dispersions are mainly used in cathodic dip coatings, as explained in Section 3.1.3, as are the epoxy resins which, with the aid of carboxylic acid groups, are converted to anionic dispersions for anodic dip coatings.

Epoxy-amine adducts bearing primary amine groups (see Equation 2.58) can also be used in traditional coatings. The amine groups are neutralised with HC_l to yield an aqueous dispersion. An epoxy resin can then be added to the dispersion, which becomes occluded in the particles of the dispersion of the epoxy-amine adduct and so is separated from the protonated amine groups outside the particle.

A pot-life of several days is then possible. After application, and evaporation of the solvent, the protonated amine and the epoxy group are in the same phase, so that a reaction to chlorohydrin (Equation 2.18) takes place. The primary amine is released and can react with other epoxy groups to build up the network. As some epoxy groups are consumed in the formation of the chlorohydrin, the use of highly-functional novolak epoxy systems (see Section 2.2.2) is recommended to ensure that enough epoxy groups are available for the crosslinking reaction [67].

Amine-functionalised (meth)acrylic copolymers [4] can also serve as curing agents for emulsified or dispersed epoxy resins.

Epoxy resins can furthermore be use to crosslink waterborne polyurethane resins. This is achieved with glycidyl ethers of glycerol (see Section 2.2.1) or polyphenol (see Equation 2.45), which are co-dispersed during synthesis of the PUR resin. As mentioned above, polyepoxide crosslinking agents can be rendered hydrophilic for use as aqueous dispersions. The resins can be dispersions, which have been stabilised with carboxylic groups or amine groups, as both of these can react with the epoxy group.

Besides PUR dispersions, water-dispersible polyolefin or polyacrylic resins can be used, either separately or in combination [101].

2.4.4 Resins for hybrids with polymers based on epoxy resins

Hybrids are resins that contain at least two different classes of polymer which are covalently bonded to each other. A representative selection from the vast number of combinations is presented below.

2.4.4.1 Epoxy acrylate

A widely employed way to introduce epoxy groups into polyacrylates utilises glycidated monomers, such as glycidyl methacrylate (GMA) or 4-hydroxybutyl acrylate (4HBAGE).

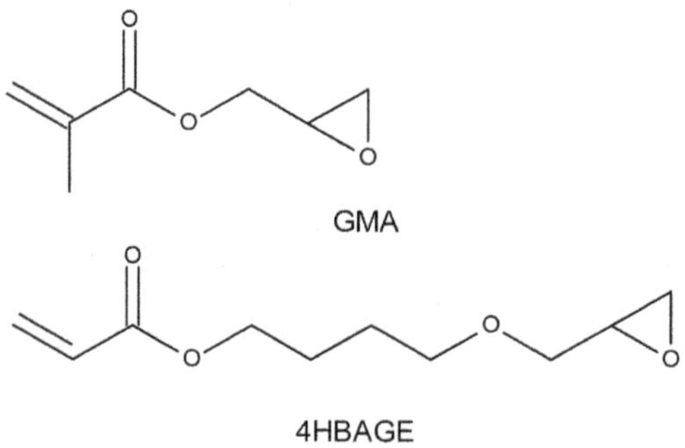

Equation 2.71: Structure of glycidyl methacrylate (GMA) and 4-hydroxybutyl acrylate (4HBAGE)

These monomers can be employed in varying concentrations as co-monomers, rendering a wide range of different performance profiles feasible [67, 101].

A further way to create dispersible polyacrylate-epoxy hybrids proceeds via the synthesis of the methacrylic ester of the epoxy resin. For this purpose, the epoxy resin is saponified with the methacrylic acid in the presence of an amine catalyst to yield the corresponding ester, followed by radical polymerisation with other, often ionic, monomers to form a water-dispersible resin [104].

Polyacrylic resins with OH-functional side-groups, obtained via ester groups of polyvalent alcohols, can be crosslinked with epoxy resins (normally low-molecular resins) [7].

It is customary to employ a 1 : 10 ratio of epoxy to acrylic, as that boosts the chemical resistance of the acrylic resin. This combination is not a hybrid in the strict sense, as the hybrid is not formed until the resins are crosslinking.

In contrast, graft polymers consisting of poly(meth)acrylates and epoxy resins are genuine hybrids. These can be synthesised in the manner shown in Equation 2.73.

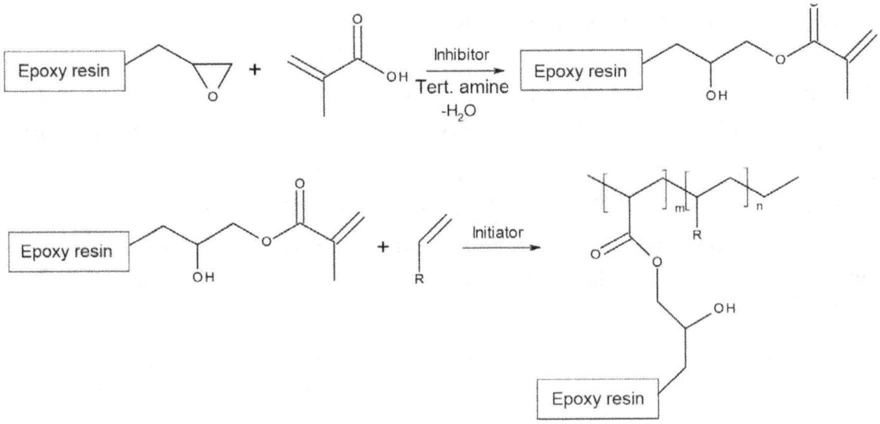

Equation 2.72: Epoxy resin/poly(meth)acrylate hybrid, according to [104]. R are typical (meth)acrylate side-groups; both cationic and anionic dispersions can be produced.

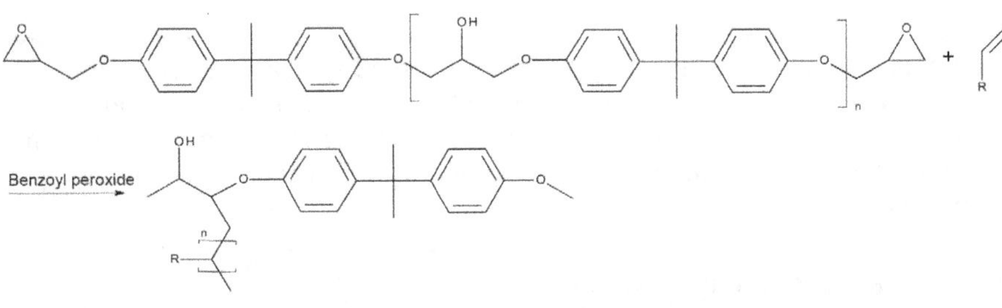

Equation 2.73: Mechanism behind graft polymerisation with an epoxy resin and (meth) acrylate monomers (cf. [4]). R are typical (meth)acrylate side-groups, with the possibility of producing cationic and anionic dispersions.

If, for example, ethyl acrylate, styrene and methacrylic acid react with benzoyl peroxide as initiator in the presence of a BPA-based epoxy resin at 130 °C in a glycol ether solvent, both benzoyloxy and phenyl radicals are formed, which can initiate the polymerisation or extract a proton from the epoxy resin. The radicals in the epoxy backbone act here as initiators, giving rise to polyacrylate side-chains on the epoxy resin. The grafting reaction happens predominantly on the aliphatic parts of the epoxy resin. The resultant solution contains epoxy resins with extended side-chains, polyacrylates and unreacted epoxy resin. The mixture is neutralised with an amine and can be converted into an aqueous dispersion. Applications for these types of resins include container coatings that are crosslinked with amine-formaldehyde resins [4, 67].

2.4.4.2 Epoxy alkyd, epoxy ester

Alkyd resins, i.e. products of the condensation of polyvalent alcohols with polybasic acids or anhydrides, can be joined to polyethers via reaction between the free hydroxyl groups and the epoxy group (see Equation 2.19) or by transesterification.

Often, low-molecular BPA-based resins are combined with oil-free or medium-oil alkyd resins based on castor oil fatty acids in a 1 : 10 ratio of epoxy to alkyd.

The epoxy resin increases the chemical resistance of the alkyd resins because the saponification-sensitive alkyd resins are enlarged due to the formation of ether groups [7].

Alternatively, BPA-based epoxy resins can be converted into epoxy esters, i.e. products of the reaction between fatty acids and epoxy resins. If hardenable fatty acids are used, the resultant resins can be cured by autoxidation.

The epoxy groups are converted into esters and hydroxyl groups via reaction with the carboxyl groups, and the hydroxyl groups can in turn be esterified (see Section 2.4.2). The reaction is done at elevated temperatures of 220 to 240 °C, usually at a conversion rate of 90 % and achieving an acid number of 7 mg/KOH.

Modifying alkyd resins with epoxy resins can greatly boost adhesion to metal substrates, and so the corrosion protection afforded by epoxy esters, compared with alkyd resin, is also markedly improved. Epoxy esters are important resins for high-grade primers and fillers in industrial coatings [14, 102], and variants produced by modification with maleic anhydride are used in anodic dip coatings (see Section 3.1.3).

2.4.4.3 Epoxy-siloxane/silicone

The addition of aminoethylaminopropyl trimethoxysilane to a BPA-based epoxy resin which has been crosslinked with a polyamide increases the heat stability, reduces yellowing at elevated temperatures, increases the hardness of the coating and improves the solvent stability due to the additional crosslinking by the siloxane [67].

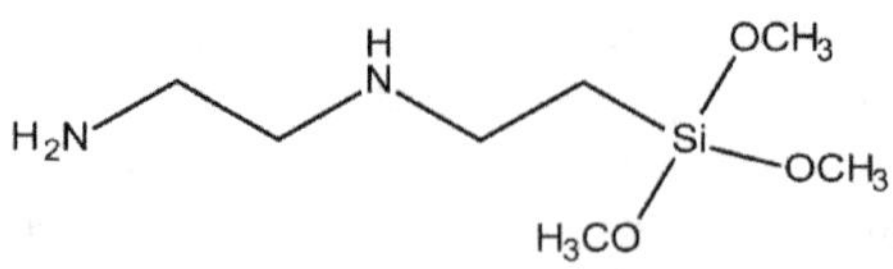

Equation 2.74: Structure of aminoethylaminopropyl trimethoxysilane

Furthermore, the addition of a siloxane in combination with a ketimine-blocked polyamine (see Equation 2.60) affords high-solids, 1-pack systems that have good storage stability [108].

In both cases, a siloxane/silicone-epoxy resin hybrid is formed.

Not only can siloxanes serve as additives for coating formulations, they can also be used to modify epoxy resins. For example, a BPA-based epoxy resin etherified with a hydroxy-functional siloxane and crosslinked with a polyamide, enhances the scratch resistance of the resultant coating and its resistance to KOH and moisture [67].

Very highly crosslinked coatings and therefore very high chemical resistance can be achieved with silicone-epoxy hybrids. First, alkoxysiloxane-containing silicones are made to react with an hydroxy-functional epoxy resin.

The resultant hybrid can be hardened with an aminosiloxane at ambient temperature; the amine groups of the crosslinking agent react with the epoxy groups, and the siloxane groups of the crosslinking agent react with the remaining siloxane groups of the hybrid, the siloxane groups having previously been hydrolysed [107].

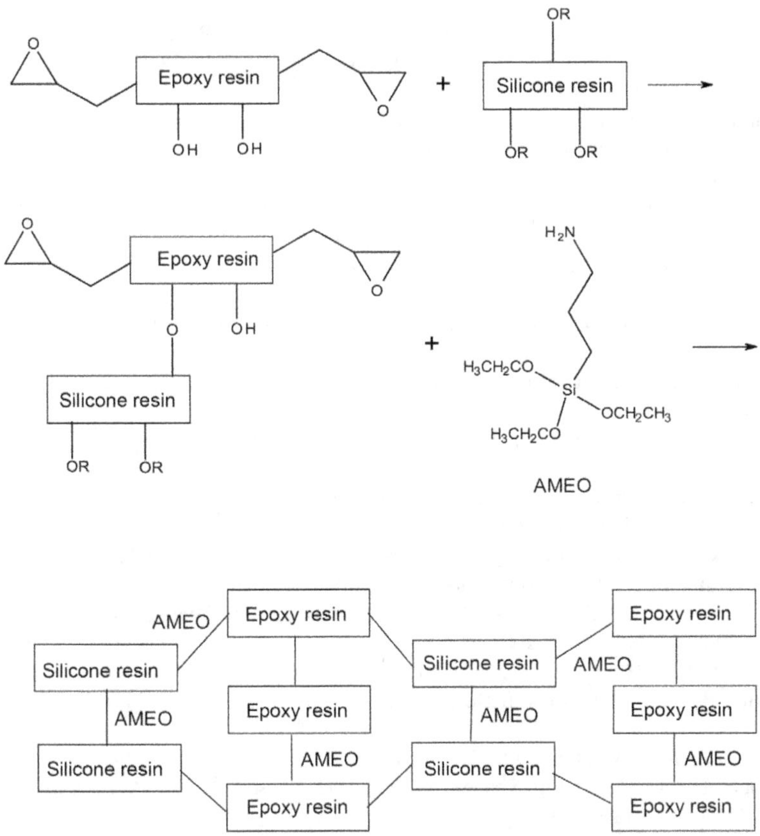

Equation 2.75: Production of a silicone-epoxy resin hybrid and crosslinking with 3-aminopropyl triethoxysilane (AMEO) [107]

2.4.4.4 Epoxy-polyamideimide

Polyamide imides are used for coating copper wires, with the solvent resistance of the resultant coating being enhanced by addition of 10 to 20 % of BPA-based epoxy resin [103].

2.5 Literature

[1] A. R. Katritzky, Handbook of Heterocyclic Chemistry, Pergamon Press, Oxford, 1985, 133–140

[2] T. L. Gilchrist, Heterocyclenchemie, VCH, Weinheim, 1995

[3] Ullmann's Encyclopedia of Industrial Chemistry, 5th Completely Revised Edition, Volume A9, Dithiocarbamic Acid to Ethanol, VCH, Weinheim, 1987, 531–563

[4] Encyclopedia of Polymer Science and Engineering, Vol. 6: Emulsion Polymerisation to Fibres, Manufacture, John Wiley & Sons, New York, 1986, 322–382

[5] R. T. Morrison, R. N. Boyd, Lehrbuch der Organischen Chemie, VCH, Weinheim, 1974, p. 623ff.

[6] Ullmanns Encyklopädie der technischen Chemie 4th Ed., Vol. 10, Dentalchemie bis Erdölverarbeitung, Verlag Chemie, Weinheim, 1975, p. 563–580

[7] H. Kittel, Lehrbuch der Lacke und Beschichtungen, Vol. 2, 2nd Ed., S. Hirzel, Stuttgart, 1998, p. 267–318

[8] Dissertation, Jan Christoph Gaukler, Einbau des Beschleunigers in Zeolithe, um die Topfzeit, Lagerstabilität und Einbrennverhalten zu optimieren, Universität des Saarlandes, 2011

[9] T. F. Saunders, M. F. Levy, J. F. Serino, Journal of Polymer Science: Part A-1 1967, 5, p. 1609–1617

[10] S. A. Zahir, Advances in Organic Coatings Science and Technology 1982, 4, p. 83–102

[11] M. D. Gilbert, N. S. Schneider, W. J. MacKnight, Macromolecules 1991, 24 (2), p. 360–369

[12] Dissertation, J. Haag, Über Dicyandiamid und eine neue daraus entstehende Base, Eberhard-Karls-Universität Tübingen, 1862

[13] Z.W. Wicks, F.N. Jones, S.P. Pappas, Organic Coatings: Science and Technology, Vol. 1, John Wiley & Sons, New York, 1992, p. 162–187

[14] A. Goldschmidt, H.-J. Streitberger, BASF Handbook Coating Technology, Vincentz Verlag, Hannover, 2007, p. 106–111

[15] J. D. Swan, Anal. Chem. 1954, 26 (5), 878–880

[16] U. Zoller, P. Sosis, Handbook of Detergents, Part F: Production, CRC Press, 2009, p. 163ff

[17] G. H. Posner, Angew. Chem. 90, 527–536 (1978)

[18] G. Zeppenfeld, L. Matejka, P. Spacek, P. Schmidt,K. Dusek, Angew. Makromol. Chem. 112 (1989) 185–194 (No. 2884)

[19] L. Shechter, J. Wynstra, R. P. Kurkjy, Ind. Eng. Chem. 49 (1957) 1107

[20] W. Tänzer, M. Fedtke, Acta Polymerica 37 (1986) No. 1, 24–28

[21] J. Klee, H.-H. Hörhold, H. Schütz, Acta Polymerica 38 (1987) No. 5, 293–299

[22] J. Klee, H.-H. Hörhold,, Acta Polymerica 40 (1989) No. 6, 421–422
[23] K. Fryauf, V. Strehmel, M. Fedtke, Polymer Bulletin 31, 183–190 (1993)
[24] Z. K. Brzozowski, G. Rokicki, B. Pogorzelska-Marciniak, Angew. Makromol. Chem. 49 (1976) 59–74 (No. 711)
[25] J. Klee, H.-H. Hörhold, H.-J. Flammersheim, Angew. Makromol. Chem. 118 (1990) 63–83 (No. 2948)
[26] D. Böschel, M.Fedtke, Angew. Makromol. Chem. 239 (1996) 201–213 (No. 4204)
[27] Dissertation, C. Wehlack, Untersuchungen zur Ausbildung der Grenzfläche Gold, Aluminium, Kupfer mit Epoxiden und PUR mit Hilfe von XPS, FTIR, Universität des Saarlandes, 2008
[28] P. Penczek, B. Szczepaniak, J. Rejdych, Acta Polymerica 42 (1991) No. 2–3, 109–112
[29] J. Klee, Makromol. Chem. 190, 2673–2681 (1989)
[30] EP000000288943A2 and therein cited patents
[31] WO002009023687A1
[32] WO002009023690A1
[33] US07282266B2
[34] J. Buddrus, Angew. Chem. 84. Vol. 1972 No. 24, 1173–1183
[35] J. Buddrus und W. Kimpenhaus, Chem. Ber. 106, 1648–1660 (1973)
[36] J. Buddrus, Chem. Ber. 107, 2050–2061 (1974)
[37] J. Buddrus, W. Kimpenhaus, Chem. Ber. 107, 2062–2078 (1974)
[38] Beilsteins Handbuch der organischen Chemie, 4th Ed., Vol. 17, Julius Springer Verlag, Berlin, 1938, p. 4f and 6f
[39] R. Dittmeyer, W. Keim, G. Kreysa, A. Oberholz, Winnacker, Küchler, Chemische Technik, Vol. 5: Organische Zwischenverbindungen, Polymere, Wiley-VCH, Weinheim, 2005, 115f
[40] Fonds der Chemischen Industrie, Informationsserie nachwachsende Rohstoffe, 2009, 51ff
[41] A. Behr, D.W. Agar, J. Jörissen, Einführung in die Technische Chemie, Spektrum Akademischer Verlag, Heidelberg, 2010, 217f
[42] B. M. Bell et al., Clean 2008, 36 (8), 657–661
[43] T. Laue, A. Plagens, Namen- und Schlagwort-Reaktionen der Organischen Chemie, B.G. Teubner Verlag, Stuttgart, 1994, 258ff
[44] K. W. Woods, P. Beak, J. Am. Chem. Soc. 1991, 113, 6281–6283
[45] Ullmanns Encyklopädie der technischen Chemie, 4th Ed., Vol. 18, Petrosulfonate bis Plutonium, Verlag Chemie, Weinheim, 1979, p. 215–217
[46] Dissertation, M. Gehring, Verhalten der endokrin wirksamen Substanz Bisphenol A bei der kommunalen Abwasserentsorgung, Technische Universität Dresden, 2004
[47] E. S. Olson, Energy & Environmental Research Center, University North Dakota, Conversion of Lignocellulosic Material to Chemicals and Fuels, Juni 2001
[48] S. Wagner, N. Graf, H. Böchzelt, H. Schnitzer, Nachwachsende Rohstoffe für die Chemische Industrie, Bundesministerium für Verkehr, Innovation und Technologie-Berichte aus Energie- und Umweltforschung, 30/2005
[49] Substantial projekt reports of "Die Fachagentur Nachwachsende Rohstoffe e.V.
[50] Shell Chemie, Technical brochure RES/CAX/1 (G), 1975
[51] Momentive, Technical data sheet, Cardura E10P, November 2012

[52] H. R. Christen, F. Vögtle, Organische Chemie, Vol. I, Otto Salle Verlag, Frankfurt, 1988, p. 620–628

[53] J. Novak, Angew. Makromol. Chem. 35 (1974) 169–176 (No. 502)

[54] W. Tänzer, H. Fiedler, M. Fedtke, Acta Polymerica 37 (1986) No. 1, 70–72

[55] H.-H. Hörhold, K. Bellstedt, D. Klemm, Acta Polyrnerica 34 (1983) Vol. 3, 135–138

[56] H.-H. Hörhold, J. Klee, H. Schütz, R. Radeglia, Angew. Makromol. Chem. 144 (1986) 1–9 (No. 2342)

[57] V. Fiala, M. Lidarik, Makromol. Chem. 154 (1972) 81–88

[58] J. Klee, H.-H. Hörhold, Makromol. Chem. 190, 3055–3060 (1989)

[59] US000004358578A

[60] US000004302574A

[61] W. A. Romanchick, J. F. Geibel, Org. Coat. Appl. Polym. Sci. Proc. 46, 410–415 (1982)

[62] M. Hesse, H. Meier, B. Zeeh, Spektroskopische Methoden in der organischen Chemie, 8th Ed., Thieme Verlag, Stuttgart, 2012

[63] K. M. Crecely et. al., J. Mol. Spectrosc. 30, 184 (1969)

[64] F. S. Mortimer, J. Mol. Spectrosc. 5, 199 (1960)

[65] J. Klee, H.-H. Hörhold, J. Raddatz, Acta. Polymerica 41 (1990) Nr11, 557–560

[66] J. Habermeier, Angew. Makromol. Chem. 35 (1974) 9–25 (No. 479)

[67] Z. W. Wicks, F. N. Jones, S. P. Pappas, D. A. Wicks, Organic Coatings 3th Edition, Wiley Verlag, 2007, 271–294

[68] Dissertation, M. Dornbusch, Synthese von fünfringheterocyclischen Retinoiden als Chromophore für das Bacteriorhodopsin, Heinrich-Heine-Universität Düsseldorf, 2001, 14–17

[69] N. R. Jagannathan, F. G. Herring, J. Polym. Sci. Part A: Polym. Chem. 26 (1988), 1–7

[70] H.-J. Streitberger, K.-F. Dössel, Automotive Paints and Coatings 2th Edition, Wiley-VCH, Weinheim, 2008, 241–244

[71] J. Ruf, Organischer Metallschutz, Vincentz Verlag, Hannover, 1993, 160–166

[72] Info brochure Deutsche Shell Chemie Gesellschaft mbH, Frankfurt am Main, I-1, Dezember 1959

[73] Info brochure Deutsche Shell Chemie Gesellschaft mbH, Frankfurt am Main, A-I-1, November 1972

[74] V. Strehmel, T. Scherzer, Eur. Polym. J. Vol. 30 No. 3 pp 361–368, 1994

[75] T. Hoang, J. Polym. Sci, Polym. Lett. Ed. 20 (1982) 417–422

[76] C. F. Poranski, W. B. Moniz, J. Coat. Technol. 49, 632, 1977, 57–61

[77] Info brochure Deutsche Shell Chemie Gesellschaft mbH, Frankfurt am Main, V-3, August 1960

[78] Info brochure Deutsche Shell Chemie Gesellschaft mbH, Frankfurt am Main, V-2, August 1960

[79] Info brochure Deutsche Shell Chemie Gesellschaft mbH, Frankfurt am Main, V-4, August 1960

[80] H. R. Christen, F. Vögtle, Organische Chemie Vol. I, Salle+Sauerländer, Frankfurt am Main, 1988, 84–85

[81] J. Ruf, Organischer Metallschutz, Vincentz Verlag, Hannover, 1993, 140f

[82] A. Goldschmidt, H.-J. Streitberger, BASF Handbook Coating Technology, Vincentz Network, Hannover, 2007, p. 81ff
[83] F. Gonzalez Garcia, B. G. Soares, Polym. Testing 22 (2003) 51–56
[84] J. D. Dorsey, G. F. Dorsey, A.C. Rutenberg, L. A. Green, Anal. Chem. 49 (1977) 1144
[85] O. Lückert, Karsten Lackrohstofftabellen, 8th Ed., Vincentz Verlag, Hannover, 1987, 461ff
[86] Safety data sheet, Aminoethylpiperazin, SysKem Chemie GmbH, Wuppertal, 2011
[87] Safety data sheet, 1,4-Phenylendiamin, Carl Roth GmbH + Co KG. Karlsruhe, 2012
[88] S. Nakano, T. Endo, J. Polym. Sci. Part A 34, 3 (1996) 475–480
[89] J. Klee, H.-H. Hörhold, W. Tänzer, M. Fedtke, Acta Polymerica 37 (1986) No. 5, 272–278
[90] J. Klee, H.-H. Hörhold, W. Tänzer, M. Fedtke, Angew. Makromol. Chem. 141 (1987) 71–81 (No. 2369) Wässrige Epoxide
[91] V. Strehmel, K. Deltschewa, K.-G. Häusler, K. Scröter, Angew. Makromol. Chem. 220 (1994) 99–109 (No. 3844)
[92] V. Strehmel, E. Zimmermann, K.-G. Häusler, M. Fedtke, Progr. Colloid Polym. Sci. 90: 206–208 (1992)
[93] K. Fryauf, V. Strehmel, M. Fedtke, Polymer, 1993, 34, 2, 323–327
[94] A. Matejicek, J. Kaska, M. Sivokova, J. Klaban, J. Horky, Angew. Makromol. Chem. 142 (1984) 1–14 (No. 2255)
[95] T. Laue, A. Plagens, Namen- und Schlagwort-Reaktionen der Organischen Chemie, B.G. Teubner Verlag, Stuttgart, 1994, 92ff.
[96] B. Müller, U. Poth, Lackformulierung und Lackrezeptur, Vincentz Verlag, Hannover, 2005, 115f
[97] A. Rybina, C. Lang, M. Wirtz, K. Grußmayer, A. Kurz, F. Maier, A. Schmitt, O. Trapp, G. Jung, D.-H. Herten, Angew. Chem. Int. Ed. 2013, 52, 6322–6325
[98] Akzo Nobel, Technical Datasheet, Thioplast G1, January 2009
[99] Dissertation, Silke Witzel, Synthese neuer funktioneller Polysulfid-Telechele und deren industrielle Applikation, Friedrich-Schiller-Universität Jena, 2007, 9ff
[100] R. Singh, G.M. Whitesides, The Chemistry of Sulphur-Containing Functional Groups, John Wiley & Sons, New York, 1993, p. 633
[101] H. Kittel, Lehrbuch der Lacke und Beschichtungen, Vol. 3, 2nd Ed., S. Hirzel, Stuttgart, 2001, p. 235f
[102] Z. W. Wicks, F. N. Jones, S. P. Pappas, D. A. Wicks, Organic Coatings 3rd Edition, Wiley Verlag, 2007, 322–325
[103] EP000000013767A1
[104] EP000000037962A2
[105] EP000000661328B1
[106] EP000002107075B1
[107] D. Hinzmann, T. Klotzbach, S. Herrwerth, FARBE UND LACK 9, 2012, 118. Jahrgang, 22–26
[108] WO002010135730A2
[109] US07300963B2
[110] US000006221934B1

3 Epoxides in coatings

Michael Dornbusch

3.1 Epoxy groups as crosslinked building blocks

3.1.1 Overview of epoxy resins and hardeners

So many resins can contain epoxy groups that it is necessary to classify the products available on the market. ISO 3673-1:1999 classifies resins by their chemical structure and possible additives, and further subdivides them into aliphatic, aromatic, and nitrogen-containing compounds.

The glycidyl ethers and esters and the epoxy groups synthesised by the Prilezhaev reaction (Equation 1.3) can therefore be considered derivatives of olefinic compounds (Table 3.1).

ISO 4597-1:2010-01 classifies hardeners for epoxy resins, distinguishing between those, which can react with epoxy groups on one hand and those, which can react with the hydroxyl groups of phenoxy resins on the other.

A striking feature is the fine distinction made for polyamide-amine hardeners, which are classified first into aliphatic and aromatic types and then by additive and filler (Table 3.2).

Remarkably, ISO 3673-1:1999 fails to classify groups of binders that do not have epoxy groups, i.e. phenoxy resins, whereas ISO 4597-1:2010-01 classifies formaldehyde resins and isocyanates as hardeners, which would suggest that phenoxy resins should be classified under epoxy resins.

3.1.2 Epoxy groups in UV-curable coating systems

Photochemical curing of the epoxy group

Direct, photochemically induced curing of epoxy groups is a cationic polymerisation, which is started by an initiator formed as a result of photolysis of a strong

Table 3.1: Classification of epoxy resins, according to ISO 3673-1:1999

Category/ Number	**I und II**	**III**	**IV**	**V**	**VI**	**VII**	**VIII**
Category/ property	Main property				Other properties		
	Chemical composition	Viscosity at 23 °C and 1 s^{-1} [Pa s]	Epoxide equivalent [g/mol]	Organic modification or solvent	Density at 23 °C [g/cm^3]	Additives	Specific information
XX	Not classified (NC)	NC	NC	NC	NC	NC	NC
01	Bisphenol A/ glycidyl ether	≤ 0.25	≤ 115	No	< 1.10	No	Material with specific combustion characteristics
02	Aromatic glycidyl ether (or -ester)	> 0.25 – 1	116 – 150	Reactive additive	1.10 – 1.14	Fillers	Amount of hydrolysable chlorine below 0.2%
03	Aliphatic glycidyl ether (or -ester)	> 1 – 5	151 – 175	Unreactive additive	1.15 – 1.19	Dyes (organic or inorganic)	Low tendency to granulate
04	Cycloaliphatic glycidyl ether (or -ester)	> 5, but liquid	176 – 210	Organic solvent	1.20 – 1.29	Fillers and dyes	Water soluble
05	Epoxide cycloolefine	semi solid	211 – 290	Reactive additive and organic solvent	1.30 – 1.39	Emulsifier	Heat resistant

Tabelle 3.1 Continue

Category/ Number	I und II	III	IV	V	VI	VII	VIII
Category/ property	Main property				Other properties		
	Chemical composition	Viscosity at 23 °C and 1 s^{-1} [Pa s]	Epoxide equivalent [g/mol]	Organic modification or solvent	Density at 23 °C [g/cm^3]	Additives	Specific information
06	Epoxide novolak	solid	291 – 525	Unreactive additive and organic solvent	1.40 – 1.59		
07	Halogenated epoxides	thixotropic	526 – 1025		1.60 – 1.80		
08	Other epoxi-dated nitrogen compounds		1026 – 2050		> 1.80		
09	Heterocyclic compounds		< 2050				
10	Epoxide-olefines						
11	Bisphenol glycidyl ether does not belong to category 1						

Table 3.2: Classification of hardeners for epoxy resins, according to ISO 4597-1:2010-01

Category	Blocks I and II	Blocks III and IV	Block V	Block VI
	Basic properties			Secondary property
	Chemical basis	**Organic modification or solvent**	**Viscosity at 23 °C and $10\ s^{-1}$ [Pa s]**	**Additives**
00	Not classified	Not classified	Not classified	Not classified
01	Not modified aliphatic polyamides	No	≤ 0.25	No
02	Modified aliphatic polyamides	Reagent	> 0.25 – 1	Filler
03	Not modified aliphatic polyamides	Not reactive additive	> 1 – 5	Colorants, organic or inorganic
04	Modified aliphatic polyamides	Solvent	> 5 – 15	Filler and colorants
05	Not modified cycloaliphatic polyamides	Accelerator	Liquid > 15	Emulsifier
06	Modified cycloaliphatic polyamides	Reagent with solvent	Semi solid	
07	Not modified polyaminoamides	Reagent with accelerator	Solid	
08	Modified polyaminoamides	Reagent with solvent and accelerator	Thixotropic	
09	Described amine cross linking agents	Not reactive additive with solvent		
10	Tertiary amines	Not reactive additive with accelerator		
11		Not reactive additive with solvent and accelerator		
12		Accelerator with solvent		

Tabelle 3.2 Continue

Category	Blocks I and II	Blocks III and IV	Block V	Block VI
	Basic properties			Secondary property
	Chemical basis	**Organic modification or solvent**	**Viscosity at 23 °C and 10 s^{-1} [Pa s]**	**Additives**
20	Condensation polymer of amine derivates with formaldehyde (urea-formaldehyde, melamine-formaldehyde etc.)			
31	Not modified aliphatic acids and anhydrides			
32	Not modified cycloaliphatic acids and anhydrides			
33	Not modified aromatic acids and anhydrides			
34	Modified acids and anhydrides			
35	Halogenated acids and anhydrides			
40	Hydrazine-derivate			
41	Dicyandiamide and derivates			
42	Boron halogene complexes			
43	Organometal complexes			
44	Onium salts			
46	Polythioles			
47	Condensation polymers of phenol-formaldehyde-types			
48	Phenole and derivates			
49	Other compounds and OH groups			
50	Free isocyanates			
51	Blocked isocyanates			
60	Ketoimines			
70	Imidazoles and derivates			

Brönsted acid. Solvent-free or high-solids systems are generally employed here. Typical photoinitiators for cationic polymerisation are onium salts of very strong acids, such as aryl diazonium salts, diaryl iodonium salts and triaryl sulphonium salts.

The resultant anion must have very low nucleophilicity, because otherwise termination of the chain by nucleophilic attack would be more likely than chain propagation as the epoxy groups are undergoing homopolymerisation.

Tetrafluoroborates, hexafluorophosphates, hexafluoroantimonate and other anions of strong Brönsted acids can be used here.

UV light releases the Brönsted acid from the initiator (Equation 3.1) to act as an acid catalyst for the homopolymerisation of the epoxy resin (see Equation 2.54), as explained in Section 2.4.2 [4, 12].

$$Ar_3S^+PF_6^- \xrightarrow{\text{UV light}} Ar_2S^{+\bullet} + Ar^\bullet + PF_6^-$$

$$Ar_2S^{+\bullet} + R{-}H \longrightarrow Ar_2SH^+ + R^\bullet$$

$$Ar_2SH^+ + PF_6^- \rightleftharpoons Ar_2S + HPF_6$$

Equation 3.1: Photolysis of triaryl sulphonium salts due to formation of a Brönsted acid Ar = aryl, R= alkyl (resin and the like) [4, 12]

For this reason, crosslinking can continue after UV irradiation has ceased, because initiation is followed by a purely thermal process.

Diaryl iodonium and triaryl sulphonium salts have very weak absorbance above 350 nm, and so excitation is only possible in the near-to-medium UV range. Photosensitisers can be used to effect excitation in the visible range. Triaryl sulphonium salts can be excited with excimer lamps at 308 nm [12].

Duration of exposure to light can also be used to control the curing reaction – polymerisation with dialkylphenylacyl sulphonium salts proceeds only upon exposure to light. The underlying mechanism utilises is the equilibrium between the ylide/Brönsted acid and the sulphonium salt (Equation 3.2).

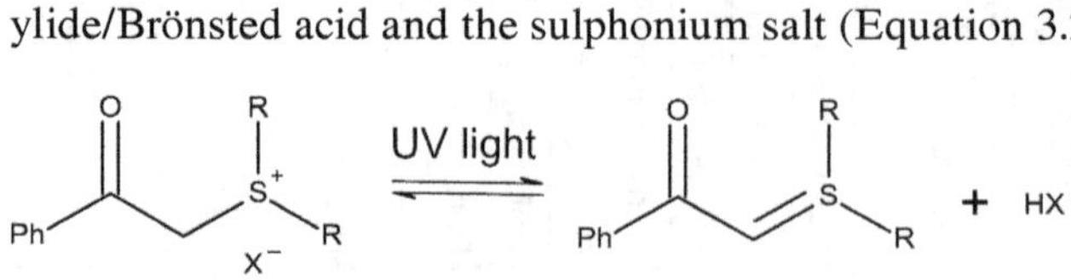

Equation 3.2: UV-induced equilibrium between ylide and sulphonium salt [4]

The equilibrium affords a way of controlling the network density.

Dialkylphenylacyl sulphonium salts are stable in epoxy resins at room temperature, and even at 150 °C, for one or two hours. Furthermore, the ylide exhibits bathochromic absorption and so photosensitisers are unnecessary [20]. In contrast to sulphonium ylides, nitrogen-containing ylides have no equilibrium and so polymerisation can continue after exposure to light and, if the right initiators are used, absorption can occur at wavelengths of up to 350 nm – again rendering photosensitisers superfluous (see Equation 3.3).

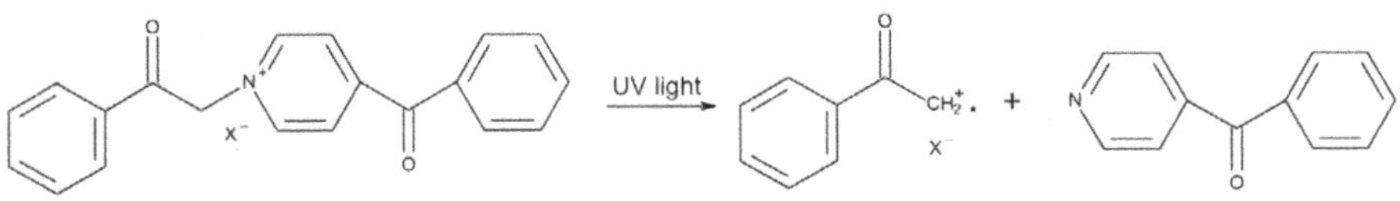

Equation 3.3: Photolysis of phenacyl-benzoyl pyridinium salts [20, 21]

In contrast to free-radical polymerisation, cations are unable to react with each other. Consequently, in the absence of nucleophilic anions, cations can initiate the crosslinking reaction. Crosslinking continues after exposure to light has ceased, provided that the cations are sufficiently mobile. It is terminated by reactive nucleophiles, such as water and alcohols, but not by oxygen.

BPA-based epoxy resins react only slowly at room temperature in conditions that favour UV-induced polymerisation. Their optimum temperature range is 70 to 80 °C.

They can be made to cure rapidly by a combination of UV and infrared radiation. Cycloaliphatic epoxies, such as 3,4-epoxycyclohexylmethyl-3',4'-epoxycyclohexane carboxylate (Table 2.7), are much more reactive than BPA-based resins. These aliphatic epoxies generally serve as thinners for cationic-curing formulations on account of their low viscosity.

Diphenyl iodonium salts have low solubility and high toxicity. Substituted derivatives, e.g. bisdodecylphenyl iodonium hexafluoroantimonate, are both more soluble and less toxic [12].

Photochemical curing of a double bond

As an alternative to direct, UV-induced reaction of the epoxy group, epoxy resins can be functionalised with double bonds that can be polymerised with photoinitiators. This can proceed in two ways: free-radical polymerisation and cationic polymerisation of the double bonds.

Free-radical polymerisation of double bonds

In free-radical polymerisation, the resins are covalently functionalised with (meth) acrylic acid or other (meth)acrylic esters or amides that contain OH-, amine or carboxyl groups.

Acrylated oligomers can be produced by reaction of the epoxy group react with acrylic acid (see Section 2.4.4) to yield esters. These resins possess outstanding chemical resistance, good adhesion and hardness. The curing reaction can involve a number of catalysts, such as triphenyl phosphine, which facilitates lower curing temperatures and thereby avoids polymerisation of the double bonds during the esterification. To make doubly sure that polymerisation does not happen, inhibitors are added to the reaction. The most commonly employed epoxy resins for this reaction are liquids based on BPA (n = 0, 1, 2, 3), and mainly yield acrylated diglycidyl ethers of BPA (Equation 3.4) [12].

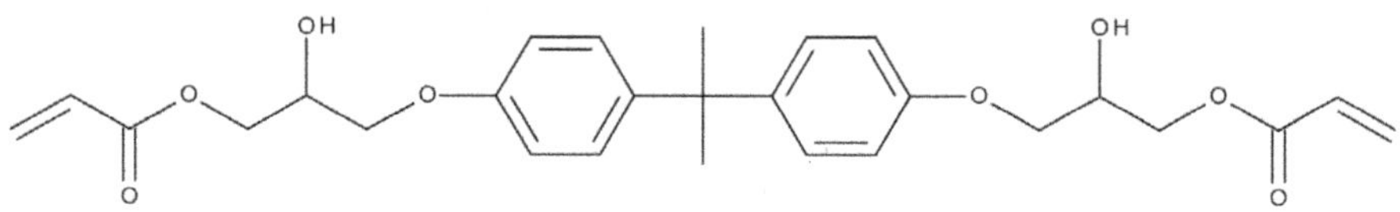

Equation 3.4: Acrylated diglycidyl ether of BPA

It has been suggested that methacrylate-containing epoxies would be suitable for enhancing the corrosion protection of printed circuit boards. The epoxy resin reacts at ambient conditions with a diisocyanate, half of which has just been made to react with 2-hydroxyethyl methacrylate. The resulting methacrylate PUR phenoxy resin improves corrosion protection extensively [9].

Cationic polymerisation of double bonds

Vinyl ethers and styrenes, especially p-alkoxystyrenes, react extremely rapidly if cationic polymerisation is initiated photochemically – this reaction is much faster than cationic polymerisation of epoxy groups. Not only is less photoinitiator needed, but the polymeric vinyl ethers are less toxic, too. Vinyl ethers, e.g. divinyl ether of BPA (Equation 3.5), serve as accelerators in epoxy resins [12].

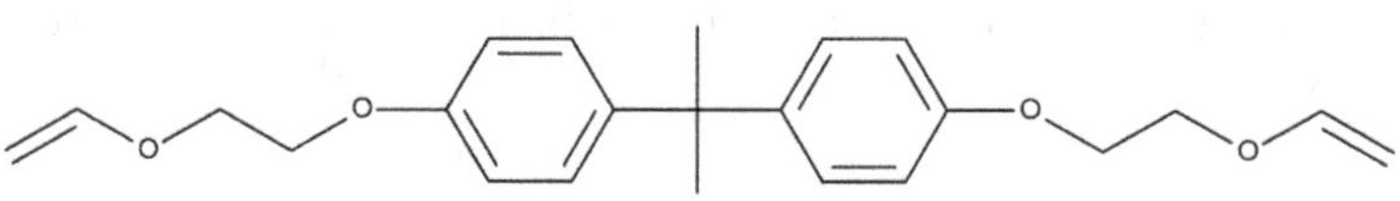

Equation 3.5: Divinyl ether of BPA produced by reaction of chloroethyl vinyl ether with bisphenol A

Combinations of curing mechanisms

With the right catalyst, it is possible to cure epoxy groups and conjugated double bonds simultaneously.

Photolysis of triaryl sulphonium salts produces not only Brönsted acids but also free-radicals. Consequently, cationic polymerisation of the epoxy group and free-radical polymerisation of the double bonds are triggered simultaneously if non-nucleophilic anions (PF6-, BF4-) are used [7, 12].

Finally, selective curing can also be performed. With diaryl iodonium tetrafluoroborate acting as initiator and the double-bond compound present in large excess (80 % methyl methacrylate) compared to the epoxy compound (20 % phenylglycidyl ether), first exposure to UV light will induce the double bonds to react and then, renewed exposure, will effect cationic polymerisation of the epoxy groups [8].

UV-curable polymers containing epoxy groups are used for coating printed circuit boards.

The UV-curable groups are cured photochemically, while the epoxy groups are thermally cured.

These so-called solder resists and protective coatings are used especially for fine-line and multilayer boards [1, 3] and are made in several steps.

The first step is to clean the circuit boards, heat them to 65 °C and then curtain coat with the solder resist. The boards can be coated homogeneously under a high feed rate without the vias in the board being filled. Coated on one or both sides, the boards are dried and exposed to UV light through a photomask. Only the regions exposed to the light are cured. The other, uncured regions are cleaned, and the solder resist is then cured at 140 °C in convection ovens [1].

3.1.3 Epoxy groups in dip-coatings

Dip-coatings can be divided into autophoretic and electrophoretic coatings, with the latter further classifiable as cathodic and anodic types.

Autophoretic coatings are deposited on the surface via a concentration gradient, which is generated close to the surface as a result of an acidic pH in the bath. By contrast, in the case of electrophoretic paints, the substrate is an electrode in an electrochemical cell that can be configured as either anode or cathode. The applied potential decomposed the water at the electrodes, raising the pH there and inducing deposition.

Electrophoretic dip-coatings

Electrophoretic dip-coatings comprise various waterborne coatings that are mainly found in primers, but are also employed in two-layer systems and pigmented one-layer systems.

Both anionic and cationic resins are used and are deposited at the anode and cathode respectively. In the case of anodic dip-coating, the negatively charged resin particles in an aqueous dispersion are deposited on the substrate, which is

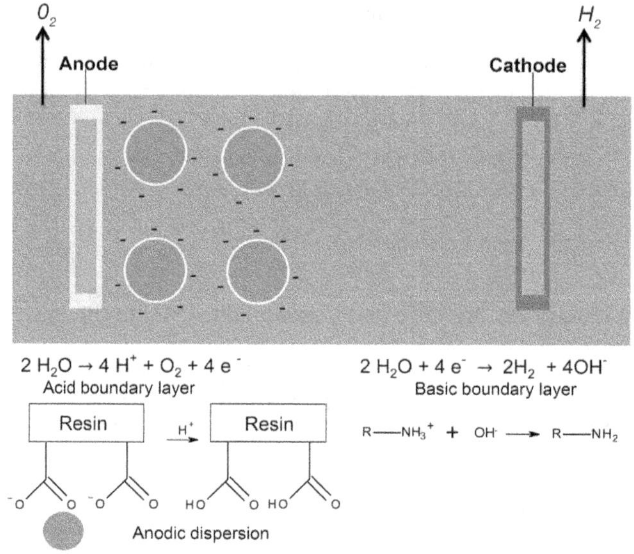

Figure 3.1: Principle behind anodic dip-coating

configured as the anode. The particles are neutralised by the positively charged protons being produced at the anode (substrate) by the electrolysis of water. The particles are destabilised and precipitate out on the substrate.

In the case of cathodic coatings, the substrate is configured as the cathode and the positively charged particles are neutralised by hydroxide ions produced by electrolysis of the water and also precipitate out on the surface [6].

In both cases, heat-curable resins are used, and all the components of the formulation must have the same electrical charge in order to precipitate in the same quantity on the substrate and to ensure that enrichment of individual components does not occur in the bath. To this end, the pigments, crosslinking agents and additives are ground or dispersed with the resin.

Anodic dip-coatings

Anodic dip-coatings were first used in the mid-1960s as primers for the automotive industry.

Initially, polybutadienes modified with maleic anhydride served as resins and were oxidatively cured with the aid of a catalyst at 160 °C. These were then replaced in the 1970s by epoxy systems. Since then, the anodic process has been

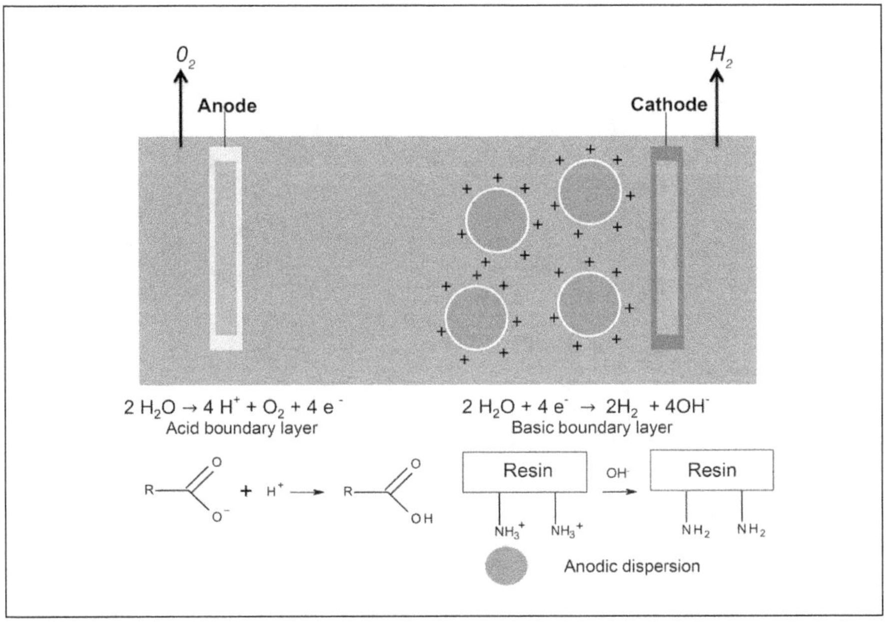

Figure 3.2: Principle behind anodic dip-coating

fully superseded by the cathodic process, because it offers better corrosion protection combined with lower coating thicknesses [2, 35].

Resins for anodic dip-coatings contain carboxyl groups with acid numbers between 40 and 80 mg KOH/g, with the pigments and all other components being ground or dispersed with the resin. The acid groups are partially neutralised with amines, such as 2-(N,N-dimethylamino) ethanol, with the result that the bath has an alkaline pH.

The first resins, based on oils modified with maleic anhydride, adhered poorly to steel and were replaced by epoxy esters (see Section 2.4.4) [10, 35].

These are produced by reaction of the BPA-based epoxy resins with unsaturated fatty acids to yield esters, which are then treated with maleic anhydride. Where the fatty acids contain conjugated double bonds, both the Ene [34] and the Diels-Alder reactions (see Equation 2.61) occur, with the Ene product capable of reacting once more with maleic anhydride in a Diels-Alder reaction.

The product of that reaction then reacts with an alcohol to generate the hemiester and is neutralised with an amine (Equation 3.6).

The resins are crosslinked with a melamine-formaldehyde resin, which is dispersed with the resin. UV-sensitive epoxy resins are unsuitable for topcoat

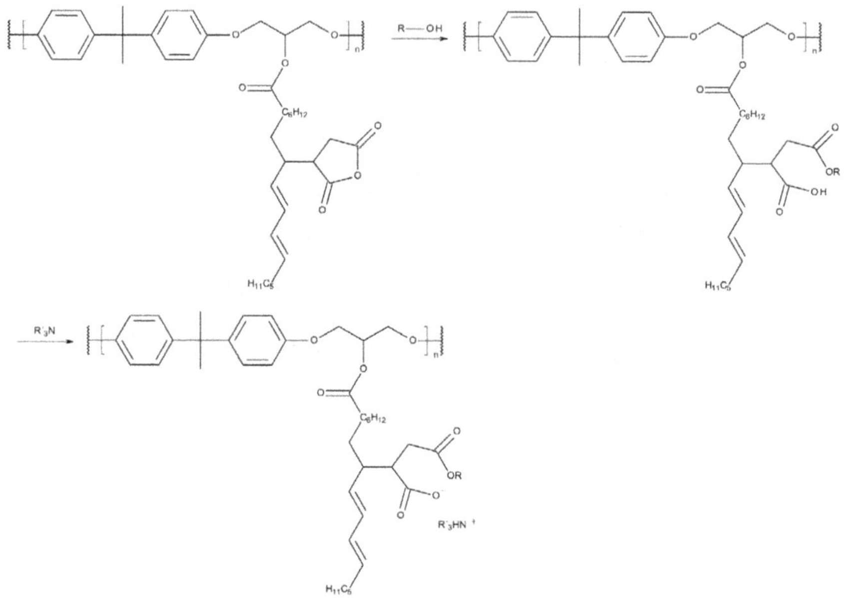

Equation 3.6: Modification of an epoxy ester with maleic anhydride (Ene product) followed by formation of the hemiester and neutralisation.

applications and so polyacrylates are used instead; these, too, are cured with melamine-formaldehyde resins [6].

A further resin for anodic dip-coatings is obtained by making a BPA-based epoxy resin react with a stoichiometric quantity of dimethylolpropionic acid. The hydroxyl-group-containing resin partially reacts with a cyclic anhydride, yielding a high quantity of carboxyl groups for neutralisation with amines (Equation 3.7).

Self-crosslinkable resins can be obtained by reaction with a semi-blocked isocyanate (Equation 3.8).

To this end, half of the bifunctional isocyanate (toluene diisocyanate, isophorone diisocyanate etc.) is blocked with compounds, such as 2-butoxyethanol, 2-ethylhexyl alcohol or ε-caprolactam, and then converted with the hydroxyl groups of the epoxy resin.

Besides electrolysing the water, anodic polarisation causes the metal substrate, usually steel, to dissolve. This gives rise to the formation of brown or yellow iron salts, which make it impossible to deposit white single-layer coatings on the substrate. Furthermore, because of the acidic pH on the anode, some of the phosphate conversion layer dissolves during the precipitation, reducing the level of corrosion protection and the adhesion of the resultant coating [6].

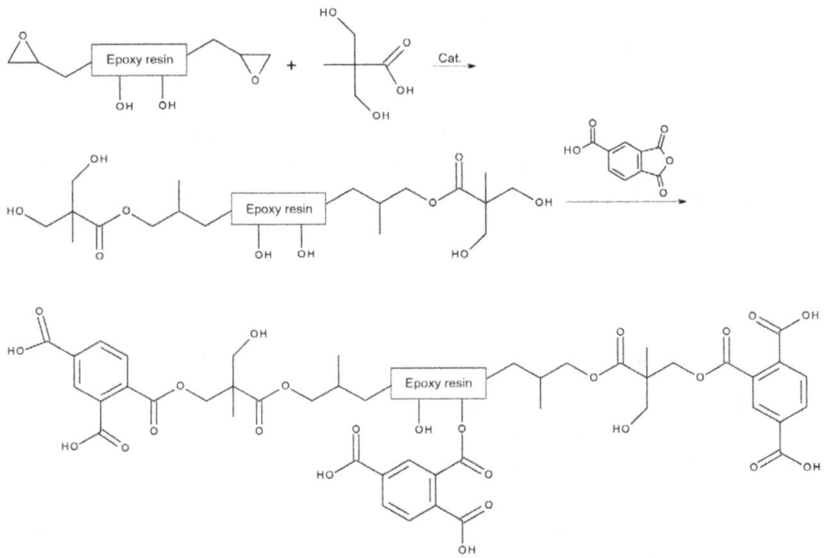

Equation 3.7: Production of a resin for an anodic dip-coating by reaction of a BPA-based epoxy resin with dimethylolpropionic acid, followed by esterification with trimellitic acid [4]

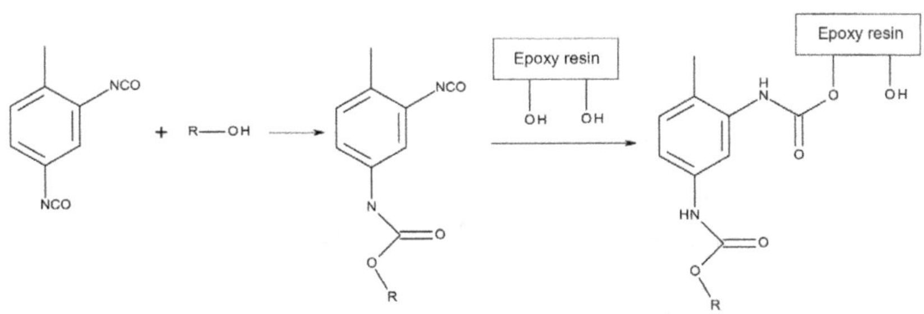

Equation 3.8: Production of semi-blocked isocyanates (alcohol acts as blocking agent), followed by reaction with the epoxy resin

Cathodic dip-coatings

The automotive industry only uses cathodic dip-coatings as corrosion protection primers. Cathodic dip-coatings were launched in the 1970s and 1980s and gradually displaced their anodic counterparts. The advantage of the cathodic process is that neither the substrate nor the phosphate conversion coating is dissolved during the deposition process because the pH at the cathode is alkaline as a result of the electrolysis of water [2, 3, 6, 35].

The coatings are based on epoxy-amine adducts (Equation 2.58) produced with BPA-based resins which are converted with hydroxylamines and then partially neutralised to yield water-dispersible resins.

Subsequent precipitation of the cationic resin occurs when the ammonium group is neutralised by the hydroxide ions produced by electrolysis of water at the cathode. Crosslinking is effected with blocked isocyanates at stoving temperatures of 170 to 210 °C. As with the anodic resins, self-crosslinkable resins can be obtained by reaction with semi-blocked isocyanates [36]. In the meanwhile, blocked polyisocyanates are now being used that are dispersed with the resin [41]. Mostly, the epoxy-amine adducts are neutralised with formic acid, acetic acid or lactic acid and so are water-dispersible. The resultant solution has an acidic pH [3, 43, 44].

Where suitable amines, such as alkanolamines or amino acid esters, are used to produce the epoxy-amine adducts, additional hydroxyl groups can be incorporated, increasing the quantity of groups available for crosslinking. However, the hydroxyl groups exert an influence on the basicity of the amine groups. For example, β-hydroxy-substituted amines are weakly basic and so require a high quantity of acid for neutralisation.

In order, nonetheless, to generate a high quantity of hydroxyl groups, strongly basic amine groups can be incorporated into the resin; this offers scope for chain extension with diols such as bisphenol A [41] (Equation 3.9).

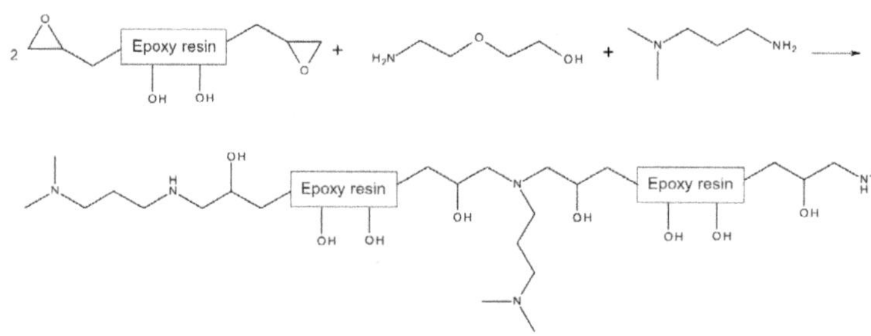

Equation 3.9: Example of the preparation of a resin for cathodic dip-coatings (see [3] and [41])

Incorporation of primary amine groups into the resin, polyamines or amino alcohols is done by conversion to ketimines (Equation 2.60) and then reaction with the epoxy resin. Subsequent hydrolysis and neutralisation release the primary, protonated amine group (Equation 3.10).

In this connection, ketimines containing thiol groups have been described that react faster with the epoxy resin (see Section 2.1.1), and so give rise to fewer side-reactions [40].

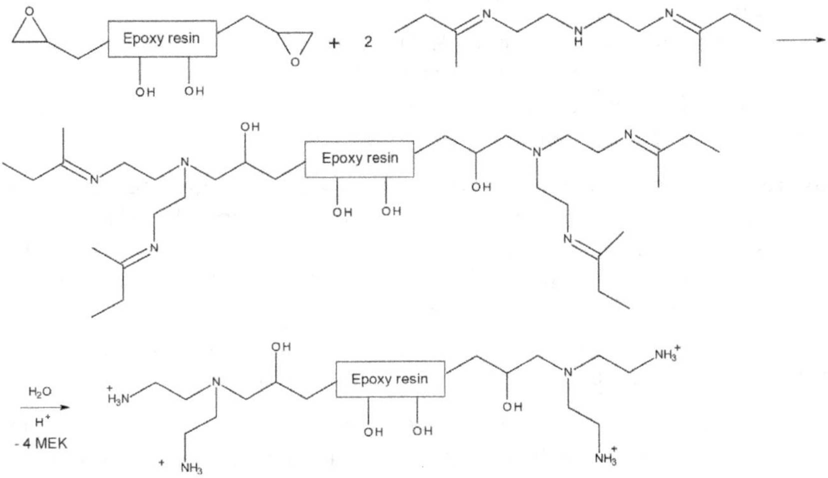

Equation 3.10: Reaction between a blocked (ketimine with methyl ethyl ketone (MEK)) polyamine and an epoxy resin, followed by hydrolysis of the ketimine and neutralisation of the primary amine group (see [40]).

Another way to incorporate amine groups utilises the Mannich reaction [45]. Here, the epoxy resin reacts with a phenol containing a ketone group. This is followed by reaction with formaldehyde and an amine:

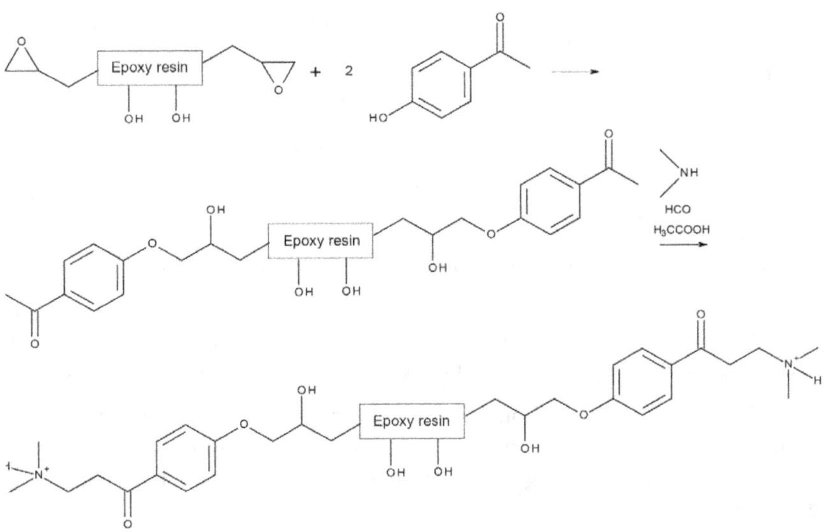

Equation 3.11: Preparation of epoxy resins containing the amine group by means of the Mannich reaction [37–39]

One alternative to epoxy resins is afforded by amine-functionalised (meth)acrylic copolymers, whereby amine-functionalised monomers, such as aminoalkyl esters and aminoalkyl amides of (meth)acrylic acid as well as vinylic heterocyclic compounds, such as N-vinylimidazole or N-vinylimidazoline, are used in the polymerisation of the (meth)acrylate.

Also used are monomers containing epoxy groups, such as glycidyl(meth)acrylate, N-glycidylacryl amide or allylglycidyl ether, which, like the BPA-based resins, are functionalised with amines or sulphonium groups. The advantage of (meth)acrylic copolymers is the high UV stability and high gloss of the resultant coating [3]. The corrosion protection properties, which are inferior to those of BPA-based resins, have improved substantially in the last few years, as a result of which these resins are widely used in pigmented one-coat systems in industry.

Not just protonated amine groups, but also other ionium compounds can serve as charge carriers. In industry, sulphonium and phosphonium groups are produced by reaction of dialkyl sulphides or phosphines with the epoxy group, usually under lactic acid catalysis. The lactic acid also serves to neutralise the resin [42].

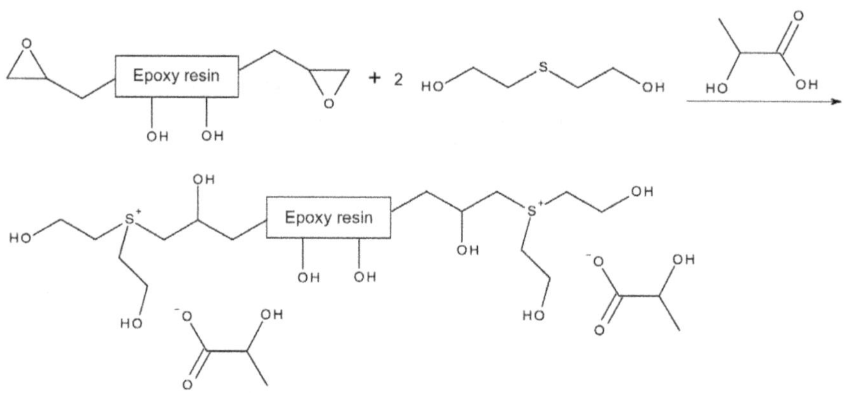

Equation 3.12: Synthesis of epoxy resins containing the sulphonium group [42]

Neutralised resins based on amines have pH values between 5 and 6.5, whereas polymers based on sulphonium or phosphonium can be stabilised in neutral conditions.

A further advantage of the sulphonium salts is that the reduction of the sulphonium group during deposition at the cathode gives rise to a polymer free-radical (Equation 3.13), thereby enabling a certain degree of crosslinking to occur as the coating is being deposited.

The pro rata incorporation (20 to 50 % of the charge carriers) of sulphonium groups benefits substrate and intercoat adhesion: the increased conductivity of the bath and the low resistance of the deposited coating facilitate high coating thicknesses while reducing throwing power by the same extent.

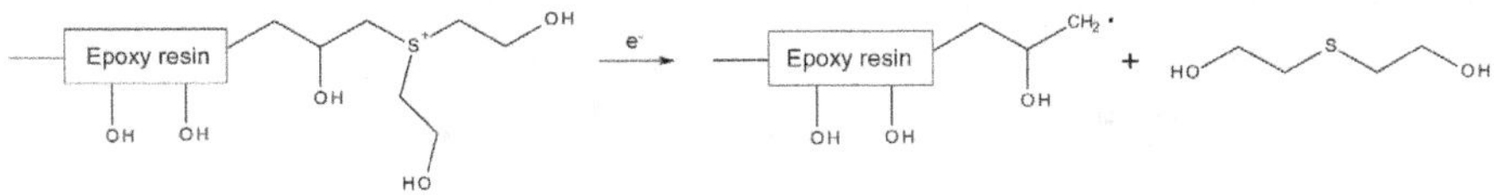

Equation 3.13: Formation of a polymer free-radical, produced at the cathode and with the potential to initiate polymerisation [3]

Finally, exclusive use of sulphonium or phosphonium groups in the cathodic dip-coating can greatly diminish yellowing of the topcoat layers [3].

Another use for resins containing sulphonium groups is that of grinding resin, especially for thick-layer coatings, because a higher potential can be applied to the substrate without fear of disruptive discharges through the coating [42].

One way to incorporate pigments into the cathodic dip-coating is to grind them with the resin (one-pack technology). The drawback is the high quantity of solvent needed to control the viscosity during grinding; 5 to 7 % of the solvent remains in the dip bath [3].

The past 15 years have seen the establishment of a technology that lowers the solvent content in the dip bath. These so-called two-pack systems contain a paste and a resin. The pigment paste is produced by grinding the pigment with a special grinding resin without solvents and then combining the paste with the resin in the bath, so that more or less no solvents remain in the dip paint.

Autophoretic dip-coatings

An autophoretic or currentless deposition process was launched on the market in 1975. This process offers the advantage of coating cavities, as there is no *Faraday* cage effect to prevent this.

Deposition involves the use of solvent-free baths ($pH < 4$) containing an organic dispersion with a solids content of 3 to 5 % iron(III)fluoride and an oxidising agent, such as hydrogen peroxide [31, 32]. When the bath is brought into contact with a steel surface, the process described in Equation 3.14 takes place.

The acid corrosion causes the concentration of iron(II) ions to rise. These ions adsorb onto and hence destabilise the dispersion, triggering deposition of the iron(II) dispersion adduct. The iron(II) ions react with the hydrogen peroxide to form iron(III) ions and the process can be run once more. The iron(II) ion concentration therefore affords a means of controlling the quantity of dispersion and thus the coating thickness.

$$Fe^0 + 2\,FeF_3 \longrightarrow 3\,Fe^{2+} + F^-$$

$$Fe^{2+} + \text{[Anionic particle]} \longrightarrow Fe^{2+}\,\text{[Anionic particle]}$$

$$2\,Fe^{2+} + H_2O_2 + 2\,H^+ \longrightarrow 2\,Fe^{3+} + H_2O$$

Equation 3.14: The deposition process for dip-coatings [32]

Coating thicknesses are usually between 13 and 25 μm; the dispersion particles in the bath have a diameter of approx. 150 nm [32].

Initially, dispersions based on polyvinylidene chloride (PVDC) and polyacrylates were used. However, it was not possible to chemically cure the layers.

This problem has been resolved by the development of BPA-based epoxy resins functionalised with dicarboxylic acids, which are then converted with semi-blocked isocyanates (Equation 3.8) [28, 31]. An alternative is to use graft copolymers based on BPA epoxy resins with different polyacrylates (see Equation 2.73) and the use of epoxy resins as crosslinking agents [29].

Further developments such as deposition on zinc phosphate steel and finally on multi-metal substrates found their first outlet in the automotive industry in Asia in 2007. Prior to this, the technology was used to coat small parts and hang-on parts, especially in the USA [30, 32, 33].

3.2 Protective coatings

Ulrich Christ

The main applications for epoxy binders are organic coatings designed for corrosion protection and surface coatings. The wide field of these applications and its enormous economic importance can be made clearer by breaking down their use into subsections. Protective coatings include anti-corrosive coatings for steel structures, coatings for shipbuilding, marine coatings, vehicle and rail vehicle construction and other industrial applications such as can and coil coatings. With a common classification for coatings there is a distinction between architectural coatings and industrial coatings according to the German Society for coatings and printing inks industry (VdL). Industrial coatings are subdivided into general industrial coatings, coatings for automotive production lines (see Section 3.1.3) and refinishing coatings, wood coatings, anti-corrosive coatings and marine coatings [49].

In the following chapter protective coatings are described in detail. It should be noted here, that powder coatings (Section 3.4) and can and coil coatings (Section

3.5) also belong to the category of industrial coatings. Protective coatings fulfil a huge range of demanding requirements, such as superior adhesion, high corrosion resistance, high chemical and mechanical resistance and in the case of exterior applications, high weather resistance.

3.2.1 Industrial coatings

Epoxy based binders are the basic part of many industrial coating systems. This product category is placed second in capacity of the entire coating production in Germany [50]. Industrial coatings are suitable for surface and corrosion protection as single or multilayer coatings for numerous consumer goods in different industrial sectors. The biggest segment here is mechanical engineering, which needs a variety of tailor-made coatings, with quite different requirements such as, for instance, food-safe, chemical and heat resistant and high performance coatings for agricultural and construction machinery [49].

Epoxy systems as single or combination binders are the materials of choice for medium to high requirements for anti-corrosive, adhesive, chemical and mechanical properties. Typical applications for epoxy binders are two component epoxy primers for agricultural and construction machinery, commercial vehicles, buses and railway vehicles.

Due to its superior anti-corrosive and mechanical properties, 2pack epoxy primers, 2pack epoxy-based intermediate coatings and even 2pack epoxy-based topcoats (dark colours) are applied to rail vehicles in the chassis and subfloor areas. Epoxy based 2pack primers are used e.g. for facade elements, the requirements of which are summed up in the GSB-guidelines (Section 3.2.2.2) [51].

The property profiles of industrial coatings are often only reached by combinations of different binders, because economic reasons need to be taken into account too. The diverse possibilities of combinations with other binder systems predestine the epoxy binder chemistry for the production of different coating technologies and for the formulation of industrial coatings with tailor-made property profiles. Figure 3.3 gives an overview of possible coating technologies and applications with epoxy groups according to [52]. Table 3.3 presents typical epoxy resin hardener combinations and application examples according to [53]. Epoxy binders are used within industrial coatings as required if improvement of adhesive strength, corrosion protection, chemical resistance and mechanical properties are needed. If corrosion protection is the main property required, epoxy systems are used as main or sole binders. Binder systems for corrosion protection with epoxy resins as main binders are described in Section 3.2.2 (2pack epoxy reactive curing systems) and Section 3.4 (epoxy powder coatings).

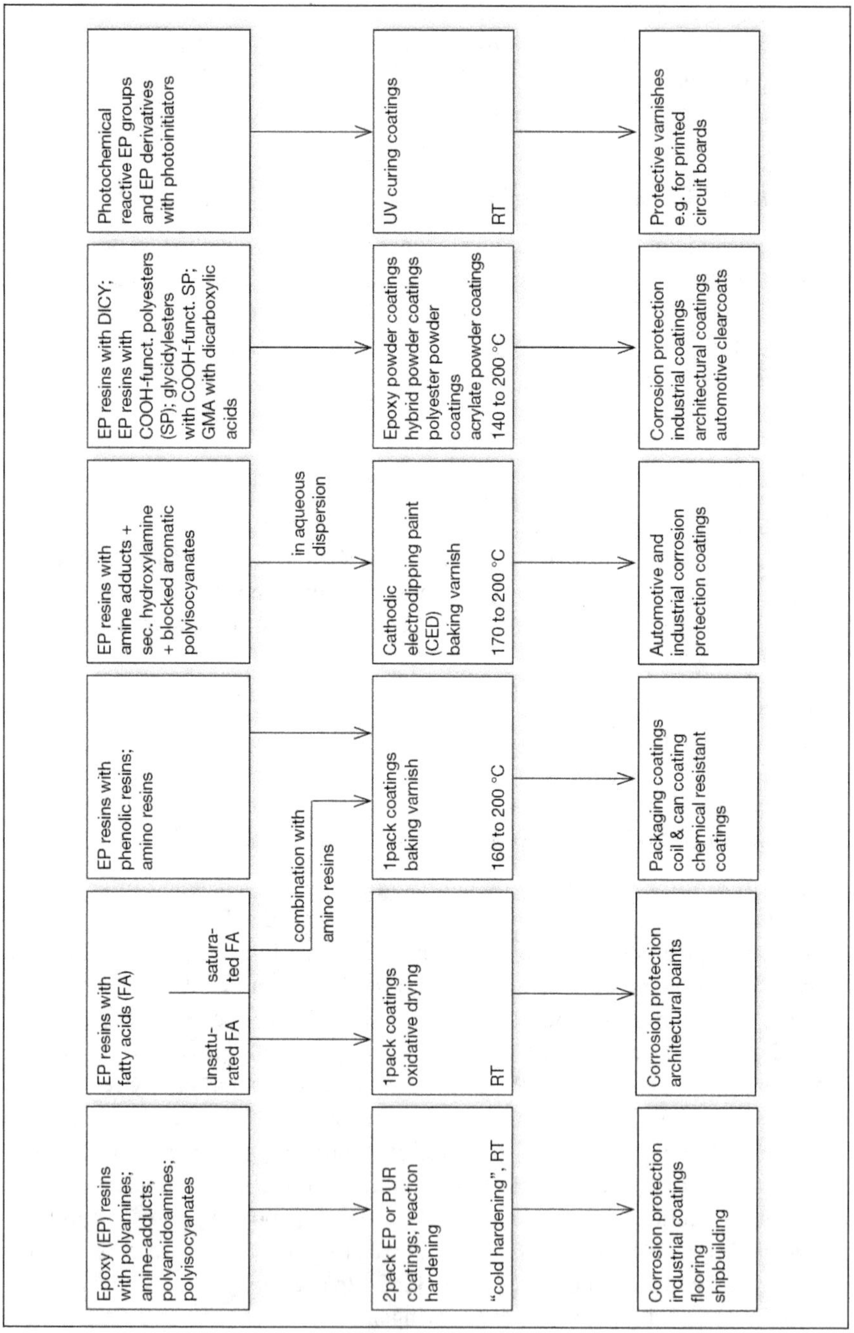

Figure 3.3: Coating technologies with epoxy groups and applications according to [52]

Table 3.3: Typical combinations of epoxy resins with hardeners for coating materials according to [53, 54]

Type of EP resin resp. other binders	Type of hardener	Crosslinking mechanism (most important)	Molecular weight [g/mol]	Typical application
Liquid	Amines and polyamidoamines	Polyaddition	≤ 400	2p EP systems for heavy duty corrosion protection
Solid, type 1 solid, type 2	Polyamidoamines & amine adducts	Polyaddition	900 1,100	2p EP Systems industrial coatings, repair coatings, marine coatings
Solid, type 3 solid, type 4 polyesters dicarboxylic acids	Latent amines, DICY, polyesters, TGIC, glycidyl esters glycidyl methacrylate	Polyaddition	1,300 1,500	Epoxide powder coatings hybrid powder coatings polyester powder coatings acrylate powder coatings
Solid, type 6 solid, type 7 solid, type 8 EPN/ECN *	Phenolic resins, amino resins, polyanhydrides	Polycondensation via OH groups; polyaddition via EP-OH groups	3,000 – 4,000	1p-cured varnishes; packaging coatings, coil coatings chemical resistant coatings
Solid, type 9	Phenolic resins, amino resins	Polycondensation via OH groups	4,000 – 5,000	Packaging and tube coatings, can coatings
Solid, type 9, thermoplast. EP resins	Isocyanates	Polyaddition via OH groups	4,000 20,000	2p PUR primers
Solid, type 4, reacted with unsaturated fatty acids	Oxygen/oxidative drying of unsaturated fatty acid esters	Oxidative drying with oxygen from air	1,500	1p-oxidative drying primers, corrosion protection, architectural coatings, single layer coatings

* EPN: Epoxy phenol novolak; ECN: Epoxy cresol novolak sole or combined with BPA resins with amine or phenolic resin hardeners are used for high chemical and thermo-resistant (high T_g) coatings due to their higher functionality compared to BPA based resins.

Combination with fatty acids – 1pack epoxy esters (air-/oxidative drying and baked coating material)

The reaction of side or terminal hydroxyl groups of fatty acids with epoxy resin leads to epoxy esters. Air dried or baked epoxy esters are produced by combining epoxy resins which are based on BPA with molar masses between 1,000 and 4,000 with fatty acids of drying oils like linseed oil or soya oil fatty acid. For each equivalent of epoxy, a minimum of 1 mole of monocarboxylic acid is used [55–57]. According to the alkyd resin used, the epoxy esters are divided into short and long oil types.

Short oil epoxy ester types usually need high-molecular weight epoxy resins with molar masses between 3,000 and 4,000; which harden well and are very durable. To produce long oil epoxy esters, epoxy resins with molar masses of 1,000 to 2,000 are used. These epoxy resins enable the production of water-dilutable epoxy esters through appropriate modifications.

The characteristics concerning the drying and hardening of these binders can be improved by additional modifications, e.g. with dimeric fatty acids, unsaturated anhydrides or with colophony. Siccatives are added to speed up the drying process. Oxidative drying epoxy ester coatings are used, amongst others, as anti-corrosive coatings (primers), boat coatings and because of their great weather resistance, as a one-coat varnish or topcoat [55, 56, 58]. Coatings based on 1pack epoxy esters fulfil special requirements, e.g. three layer paint structures with a 240 µm layer thickness on blasted and zinc coated steel substrates, which meets the requirements of ISO 12944-6 of corrosivity category C 4 high [59].

Epoxy ester baked finishes are made through the reaction of epoxy resins based on BPA with molar masses between 1,500 and 4,000 with non-drying fatty acids like castor oil fatty acid or drying fatty acids like tall oil fatty acids as well as with amino resins. Depending on the degree of esterification, short, middle or long oil epoxy esters are available. Through these modifications, one component epoxy esters are used as binders for primers, one-coat or multi-coat paint systems. For primers, mostly urea resins and for topcoats melamine resins are used. In these formulations, the weight percent ester to amino resin lies in the range of 90:10 to 70:30. These systems are hardened at baking temperatures between 120 and 180 °C. The appropriate coatings show, in addition to the characteristics of alkyd resins like good weather resistance and minor yellowing, the following features: an improved adhesive strength, increased corrosion protection, enhanced chemical resistance, and a higher mechanical stability. The latter characteristics are based on the resulting ether function through the chemical reaction with the epoxy component. The short and middle oil epoxy esters are used mainly for baked varnishes with high durability and good chemical resistance [55, 58, 56].

Combination with isocyanates – fast and low temperature hardening 1pack or 2pack-polyurethane (PUR) coatings

In chemical reactions of epoxy resins with di- or polyisocyanates to get polyurethanes, mainly higher-molecular epoxy resins, based on BPA, are used with molar masses higher than 4,000, or appropriate phenoxy resins with a sufficient amount of cross linking hydroxyl groups. Aromatic polyisocyanates based on toluene diisocyanates (TDI) are mostly used as hardener, because of the higher reactivity and lower price compared to aliphatic isocyanates.

Hardening with isocyanates as compared to hardening with polyamines has the big advantage of faster curing at low temperatures down to approx. 0 °C and producing a very high durability coating against solvents and acids and good durability against bases.

The disadvantage of hardening epoxy resins with polyisocyanates is a higher sensitivity to use in situations when the surface is in contact with water. Alternatively, 2pack coating materials hardened by isocyanates are used amongst others for anticorrosive coatings for storage tanks within the chemical industry as well as coatings in the field of wastewater. The advantage of fast drying is used, amongst others, for corrosion protection, e.g. in shipbuilding and as shop primers [60, 61].

Blocked isocyanates in mixing ratio with one hydroxyl equivalent for 0.4 to 0.9 isocyanates equivalents are used for the formulation of baked coatings [62]. They are cured at approximately 150 °C and show coating properties with high corrosion protection, high durability against chemicals, high flexibility and high hardness. The films of isocyanate hardened packaging coatings are usually more flexible but less resistant against chemicals than amino resin hardened systems [58, 63].

Combination with amino resins – baked coating materials

For the chemical reaction of the hydroxyl groups of epoxy resins with amino resins, i.e. melamine formaldehyde, urea formaldehyde or benzoguanamine formaldehyde resins, usually higher-molecular weight solid resins such as BPA with epoxide equivalents of 1,500 to 3,500 or phenoxy resins are combined partly or totally with butanol, methanol or isopropanol etherified amino resins. According to the mixing ratio, epoxy to amino resin, the properties of the coating film can be adjusted. A higher EP resin proportion usually improves adhesion and elasticity and a higher amino resin content increases the resistance against chemicals and solvents. Urea resins are used with a mass ratio between EP resin and amino resin of approx. 75 : 25 and 60 : 40. Melamine and benzoguanamine combinations are used between 80 : 20 and 70 : 30 [64]. Curing temperatures of such formulated paints are between 150 and 200 °C dependent on additive catalysts such as p-toluene sulphonic acid. Coatings from combinations of amino and epoxy resins show

Table 3.4: Guide formulation of a water-dilutable 1pack baked coating material [65]

Pos.	Raw material/generic name/trade name	Parts by weight
01	Epoxy-BPA solid type 7	**88.40**
02	Wetting agent/defoamer	**0.80**
03	Defoamer ester alcohol	**0.80**
04	Acid catalyst, amine blocked	**0.50**
05	Levelling agent	**0.30**
06	Corrosion inhibitor	**0.45**
07	Additive for substrate wetting and levelling	**0.45**
08	Melamine resin	**8.30**
Sum		**100.00**
01	e.g. "Beckopox" EP 2307w/45WAMP (Allnex)	
02	e.g. "Syrfinol" SE-F(Air Procucts)	
03	e.g. "Texanol" (Eastman)	
04	e.g. "CYCAT" VXK 6395 (Allnex)	
05	e.g. "Additol" XW 395 (Allnex)	
06	e.g. "Anticor" AM 4501(Addapt Chemicals)	
07	e.g. "Additol" VXW 6503(Allnex)	
08	e.g. "Cymel" 303 LF resin (Allnex)	
Characteristics of coating material and cured coating film		
Baking conditions		**160 °C/20 min**
Non volatile contents		approx. 49 %
Dynamical viscosity DIN EN ISO 3219, 23 °C		approx. 7.000 mPas
Application on steel, CRS; dry film thickness		50 µm
Pendelum hardness acc. König DIN EN ISO 1522		approx. 220 s
Chemical resistance/resistance to MEK: number of cycles without dissolution of coating surface		1.000
Adhesion: Cross-cut characteristics DIN EN ISO 2409		Gt 0–1

high adhesion, excellent impact resistance, good chemical resistance, high surface hardness and good abrasion resistance as well as high flexibility. Urea resins are preferred, e.g. for usage in primers and one-coat systems, amongst others, for packaging coatings if the optimization of the adhesion of the coating is most important. Combinations with melamine resins are best used in topcoats, because of their good chemical resistance, good gloss retention, and resistance against yellowing [58, 62]. Table 3.4 shows a guide formulation of a 1pack baked coating material containing a binder system consisting of a combination of BPA epoxy resin, type 7, with a melamine resin. Resulting coatings show high chemical and solvent resistance [65].

Combination with phenolic resins – baked coating materials

Higher-molecular weight epoxy resins on a BPA basis (molar mass > 4,000) or phenoxy resins are preferably used in reaction with alkyl phenols or BPA based resoles or mostly with butanol etherified hydroxymethyl group derivatives. The chemical reactions are carried out as for amino resins at 160 to 200 °C, enhanced by acidic catalysis.

The properties of coatings based on these binders are dependent on the mixing ratio of epoxy to phenolic resins. A higher epoxy proportion usually improves the adhesion and elasticity, whereas a higher phenolic proportion leads to a better chemical resistance of the coating. The mass ratios of epoxy and phenolic resins are preferably in the proportion of 80 to 20 % for more flexible and at 60 to 40 % for higher chemical resistant films. Baked coatings based on these binders show an even higher chemical resistance than epoxy amino resins. They are used in very thin layers on the inside of food cans and tubes, for example, against fruit acids, because of their excellent adhesion and their high corrosion and chemical resistance. Their strong tendency to yellowing is tolerated as the, so called, gold varnishes for inside can coatings.

Additionally, anhydride and carboxyl terminated low-molecular weight polyester resins are used as crosslinkers for white none yellowing coatings for the inside of cans [56]. Furthermore, epoxy phenolic resin baked coating materials are utilized for white goods as well as for wire and electrical insulation [66].

3.2.2 Corrosion protection

Corrosion protection through anti-corrosive coatings

The principle of corrosion protection by the use of organic coatings is the separation of the substrate from the attacking (corrosive) medium. The special significance of this technology is shown through the fact, that 80 % of all surfaces which should be protected against corrosion are protected with such organic coatings [67] and that this principle of anti-corrosive coatings for steel constructions has been successfully used for many decades. ISO 12944 [68] describes the essential characteristics which are important for an appropriate corrosion protection system for steel constructions. These are, amongst others, environmental conditions and the duration of protection, which concerns the kind of substrate surface pretreatment, the selection, the composition and testing of the coating systems.

The corrosion protection with organic coatings is usually realized by the application of a multi-layer system, as shown in Figure 3.4. This system normally consists of a primer layer, (priming coat (PC)), one or more intermediate layers (intermediate coats (IC)) and a topcoat layer (topcoat (TC)).

The aim of the pretreatment is to eliminate extraneous matter from the metal surface to provide an inert surface, which has good stability and will provide excellent adhesion characteristics to the subsequently applied organic coating [69].

The priming coat provides the protection against corrosion of the substrate and ensures a stable adhesive bonding with the substrate material and the intermediate coat.

The prerequisites for good corrosion protection, are good (wet) adhesion of the binder system of the priming coat on the metal substrate, the anti-corrosive properties of the active anti-corrosion pigments as well as a barrier function against the transport of water, electrolytes and oxygen into the interface of coating and metal substrate.

The intermediate coating(s) support (s) the barrier function through its high layer thickness and lamellar pigments or fillers. Additionally, the intermediate coating has a surfacing function and equalizes irregularities in thickness of the primer and irregularities on the metal substrate. The number of intermediate layers is dependent on the requirements for the coating system.

The topcoat protects the whole coating film against influences of weather (UV-rays), against harmful substances, chemicals and mechanical stress. The topcoat also has coloristic functions, such as colouring, brilliance and other surface effects.

In the field of corrosion protection of steel structures, so-called duplex systems are very important. They consist of a metal coating on the substrate to be protected, like hot dipped galvanizing of steel surfaces in combination with organic coatings [70].

Composition of 2pack epoxy anti-corrosive coatings

The corrosion protection properties of the coating system are provided by anti-corrosion pigments, zinc dust and barrier pigments, such as micaceous iron oxide. However, it is crucial that the coating formulation is optimised to ensure compatibility with these components and the binder. This requires the appropriate selection and testing of all component materials.

The impact of active anti-corrosion pigments and especially within water-dilutable coatings and their specific interaction with the binders has been the topic of comprehensive studies for many years. Zinc phosphates have gained high importance as active anti-corrosion pigments [71–75] and the range of applications increased due to improvements through modification of zinc orthophosphates and

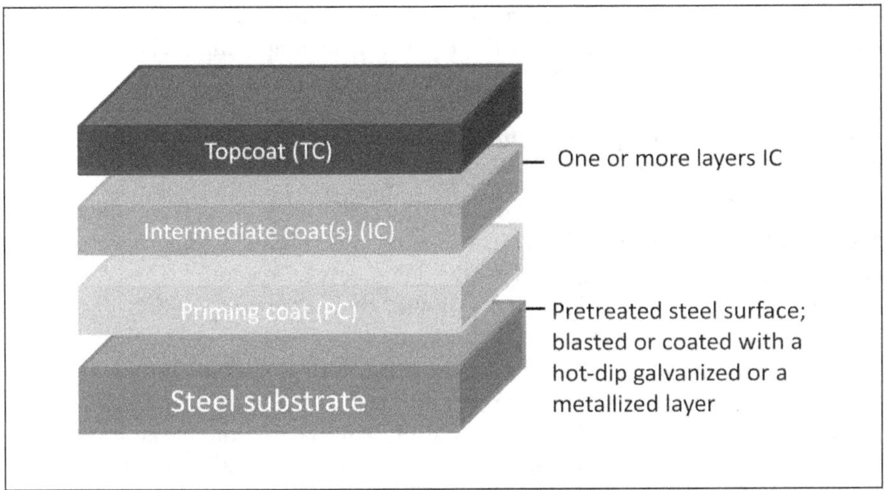

Figure 3.4: Multi-layer system of organic coatings for corrosion protection

polyphosphates [76]. Investigations have been carried out concerning corrosion protection of the so-called third generation of anti-corrosion pigments like modified zinc aluminium polyphosphate, zinc free strontium aluminium polyphosphate and zinc phosphate modified with zinc molybdate in contrast to zinc phosphate. A clearly reduced undercreepage under salt spray testing when compared to zinc phosphate was shown within solventborne 2pack epoxy coatings on steel [77–79]. For example, it is assumed that the reason for this is the precipitation of a protective layer on the steel surface that limits the influx of aggressive species onto the steel surface and restricts the cathodic delamination [77, 78]. Latest developments of zinc free active anti-corrosion pigments are, for example, based on magnesium modified calcium phosphate [76, 80] or based on calcium strontium phosphosilicate and dicalcium phosphate [81].

The effect of barrier pigments like micaceous iron oxide for the uptake of water and the water balance of anti-corrosive coatings such as described in ISO 12944 [68] is discussed in the literature and is the item of further investigations [82, 83]. In these studies, micaceous iron oxide reveals both a reduction [84] and an increase of water uptake in anti-corrosive coatings [85].

The specific properties of anti-corrosive coatings resulting from the use of certain raw material types are given in ISO 12944 [68] and TL/TP-/KOR [86] (see Section 3.2.2.2). They give detailed recommendations on the composition of the coating material, e.g. the technology of the epoxy resin and hardener, type and concentration of active anti-corrosion pigments, amount of zinc dust pigments, as well as concentration of micaceous iron oxide pigment.

These recommendations for the coating material composition and the extent of tests and inspection are shown in the following section using a chosen coating material from appendix E of TL/TP-/KOR [86].

Guidelines for the composition and testing of 2pack coating materials according to TL/TP-KOR

As shown in the following Section 3.2.2.2, TL/TP-KOR describes in Section 5 "Testing of coating materials" and lists coating systems and coating structures for appropriate corrosivity categories. In the TL sheets of appendix E, coatings and coating structures are described in detail with data about composition, characteristic values and testing methods for final quality control of liquid coating material, characteristic values of properties in end use requirements, processing and dry film (applied and cured coating) condition. Following, an example is shown taken from TL sheet 87 "EP/PUR" "Coating materials on the basis of epoxy or polyurethane (high-solid coating materials)". With this coating system, in general, two different coating structures are possible:

Coating structure	Corrosivity category/ durability range
System 1 PC: EP zinc phosphate IC: EP micaceous iron oxide (MIO) TC: PUR with or without MIO	C5 I, high, C5 M, high
System 2 PC: EP zinc dust IC: EP micaceous iron oxide (MIO) TC: PUR with or without MIO	> C5 I, high, C5 M, high (> 25 years)

Table 3.5 lists data concerning composition, characteristic values and testing methods for final quality control of liquid coating material, processing, and dry film condition of one coating material from appendix TL-E sheet 87 "EP/PUR" [87]. The data is for one colour shade of the priming coat, intermediate and topcoat material from TL sheet 87 ("EP/PUR").

The most important applications of epoxy resins within coatings are the industry segments protective coatings and shipbuilding [58]. Part of these industry segments are the application fields of steelwork and steel structural engineering including bridges, electrical towers, chemical facilities and tank farms, power plants, pipes, signal facilities for road and rail traffic, overseas containers, steel construction for hydraulic engineering and offshore facilities [88].

A number of factors need to be considered when formulating a protective coating. These include the environmental conditions to which it is exposed, the re-

Table 3.5: Coating materials based on EP/PUR (extract from appendix E sheet 87 "EP/PUR" [87]*)*

Coating materials based on EP and PUR			**Sheet 87 EP/PUR**	
1 Paint Composition in delivery state Tests according to annex D No. 1				
Material-No.	Color RAL or description	Composition of pigments based on 100 wt % pigments	Wt-% part	
			Binder	Solvent
1.1 Coating materials for primer coatings (PC)				
1.1.1 Coating material for first EP-primer coating				
687.02*	Sand yellow RAL 1002	20–25 % zinc phosphate or modifide zinc phosphate 10–15 % zinc oxide, remainder: fillers and pigments for tinting	17–22	≤ 30
687.03*	grey	≥ 94 % zinc dust, remainder: fillers	7–10	≤ 20
1.2 Coating materials with micaceous iron oxide (MIOX) for intermediate coating (IC) or top coat based on EP				
687.12*	grey DB 702	≥ 75 % MIOX, remainder: fillers and pigments for tinting	18–24	≤ 32
1.3 Coating materials for topcoats (TC)				
1.3.1 Coating materials with micaceous iron oxide (MIOX) or topcoats based on PUR				
687.52	blue DB 501	≥ 65 % MIOX remainder: fillers and pigments for tinting	24–30	≤ 35

* These coating systems of the respective TL-data sheet have to be tested in general

Tabelle 3.5 Continue

1.4 Binders					
Coating materials	Master component		Hardeners		Testing methods
EP coating materials	**Cold hardening epoxy resins epoxy resin equivalent ≥ 400**		Polyamine adducts and/or polyamine and/ or polyaminoamides or polyaminoamide adducts and other components necessary for hardening, **amin number ≤ 220** DIN 16 945		IR-spectroscopy Annex D No. 2
PUR coating materials	Hydroxyl group containing acrylate resins		Polyfunctional aliphatic isocyanates DIN 16 945, synthesis via chemical reaction of hexamethylendiisocyanate (HMDI); HMDI content < 0,5 %		
2 Characteristic values and properties for final quality control of paints (delivery state)					
3 Characteristic values and properties for processing the coating materials					
3.1 Curing time	Coating material	Drying stage	Climate conditions		Dry film thicknesses: Material No. 687.03: **70 µm** other materials: **80 µm** DIN EN ISO 1517 DIN 53 150
			NK 23/50 DIN EN 23 270	7 °C/85 % rel. humidity	
	Material-No. 687.02/06	1	≤ 2 h	≤ 3 h	
		6	≤ 16 h	≤ 48 h	
	Material-No. 687.03	1	≤ 1 h	≤ 2 h	
		6	≤ 8 h	≤ 16 h	
Test/properties	Requirements/values		Test methods		
3.3 Sagging behaviour	Dry film thickness at Mat-No. 687.03: **80 µm** Other Mat-No.: **120 µm**		Annex D No. 5		

Tabelle 3.5 Continue

Test/properties	Requirements/values		Test methods
3. Characteristic values and properties for processing the coating materials (continuation)			
3.4 Processing time (potlife)	EP coating materials: ≥ **8 h** PUR coating materials: ≥ **6 h**		Annex D No. 6
3.5 Overpaintability	Possible after curing time of: ≤ **16 h**		Dry film thickness: **80 µm** testing according Annex D No. 7
4. Characteristic values and properties of coating in dry film state			
Test/properties	Requirements/values		Test methods
4.1 Resistance to humidity (continuous condensation) ISO 6270-1	Blistering: rusting: cracking: flaking: adhesion cross-cut:	0/0 Ri0 0/0 0/0 ≤Gt1	1. Test samples: system 1 and 2: steel panel 100 x 150 mm 2. Surface pretreatment system 1 and 2: sandblasted, Sa 2½ DIN EN(ISO 8501-1), roughness "medium (G)" DIN EN(ISO 8503-1) 3. Coating structure and dry film thickness System 1: 1 x **PC** Mat.-No. 687.02: **80 µm** 1 x **IC** Mat.-No. 687.12: **80 µm** 1 x **TC** Mat.-No. 687.51: **80 µm** System 2: 1 x **PC** Mat.-No. 687.03: **70 µm** 1 x **IC** Mat.-No. 687.12: **80 µm** 1 x **TC** Mat.-No. 687.51: **80 µm** 4. Testing according Annex D No. 9 **test durations:** system 1: **720 h** system 2: **1200 h**

Tabelle 3.5 Continue

4.2 Resistance to humidity in presence of sulfur dioxide	Blistering: cracking: flaking: adhesion cross-cut:	0/0 0/0 0/0 ≤Gt1	1. Testing samples, surface pretreatment, coating structure and curing time and dry film thickness as Section 4.1, system 1 and system 2 2. Testing according Annex D No. 10 Test duration: **30 cycles**
4.3 Resistance to salt spray fog ISO 9227 NSS	Undercreepage at scribe: blistering: rusting: cracking: flaking: adhesion cross-cut:	≤3 mm 0/0 Ri0 0/0 0/0 ≤Gt1	1. Testing samples, surface pretreatment, coating structure and curing time and dry film thickness as chapter 4.1, system 1 and system 2 2. Testing according Annex D No. 13 3. **Test duration:** system 1: **1440 h** system 2: **2160 h (with PC with Zinc dust)**
4.4 Coating structure 1 (Intercoat adhesion after outdoor weathering of primer coating)	No loss of adhesion		
4.5 Coating structure 2 (Intercoat adhesion after outdoor weathering of intermediate coating)	No loss of adhesion		
4.6 Artificial weathering			
4.7 Long time performance	Blistering: rusting: cracking: flaking: adhesion cross-cut:	0/0 Ri0 0/0 0/0 ≤Gt1	1. Testing samples, surface pretreatment, coating structure and curing time and dry film thickness as Section 4.1 2. Testing according Annex D No. 16 **test duration: 24 months**

quired duration of the corrosion protection and the decorative properties of the cured coating. Other factors to be considered are environmental protection and the cost of the coating material. To obtain the highest standards of corrosion protection and the flexibility to apply the coating under a wide range of environmental conditions (temperature, relative humidity, air movement) normally requires cold curing i.e. reaction curing coating systems based on 2pack-epoxy and 2pack-PUR systems would be considered [58].

Standard 2pack epoxy primer coating materials

Partly solventborne standard epoxy primer materials such as epoxy resins based on BPA, solid resin type 1 or 2 with polyamidoamines (PAA) hardeners are used in certain fields where outstanding corrosion protection is a must and no restrictions concerning solvents are given, for example, in shipbuilding within shipyards and on buildings outdoors. PAA hardeners have good compatibility with epoxy solid resins, enable a fast curing and ensure a good corrosion protection. A catalyst is needed for curing at lower temperatures. Standard 2pack epoxy systems have nonvolatile fractions from 40 to 60 mass-% or approx. 50 vol-% equivalent to 400 to 600 g/l VOC. These products proved their worth for more than 50 years because they included high-molecular weight resins and hardeners as anti-corrosive coating systems with good adhesion and good physical properties. If good adhesion and flexibility of the coating are needed, crosslinking with PAA hardeners is preferred. To obtain a better compatibility of resin and hardener a pre-reaction period is required. The usage of PAA epoxy pre-adducts need, in contrast, no pre-reaction period. For 2pack epoxy coatings with a high resistance against chemicals, low-molecular weight epoxy resins based on BPA solid resins are combined with epoxidized novolaks (EPN: epoxy phenol novolak). This leads to higher degrees of crosslinking because of their higher functionalities [89, 90].

For so-called functional coatings with high chemical and mechanical resistance, amine adducts are mainly used either in-situ or as isolated products. Such coating materials have shorter pot lives and cure faster to films with lower adhesion and lower flexibility [61].

Due to more stringent environmental legislation, more and more low-solvent 2pack epoxy coating materials are being used. A big challenge here is to completely substitute proven solventborne systems in highly corrosive environmental conditions [88, 91].

Environmentally friendly 2pack corrosion protection systems

The intentions of the German paint industry together with raw materials suppliers, paint manufacturers and paint users is to act in a responsible way. This concerns environmental protection by conforming to European as well as national environmental and health and safety legislation, aimed at significantly reducing solvent consumption during production and usage of coating materials [92, 93]. This includes new product developments to provide coatings with thinner layer thicknesses and coating systems with fewer layers to provide significantly improved corrosion protection than is currently being achieved.

Additionally, it is intended to reduce the consumption of materials and energy as well as the costs for application and pre-treatment of surfaces without any loss of corrosion protection. With the implementation of the new EU-VOC guideline [94, 95], the development and the introduction of coating materials with low solvent content, was further encouraged. This led to the development of coating materials, which are based on different coating technologies, like high-solid and water-dilutable 2pack products or powder coatings.

2pack epoxy coating materials, which fulfil the requirements of the VOC guideline [94] concerning application of coating materials within stationary facilities with a VOC threshold of 250 g/l will be described in the following section. To adapt the coating materials to the new VOC guideline required the development of epoxy resin and hardener technologies with a significant decrease in solvent content, and on the other hand, the development of solvent-free systems with water as solvent or dispersant. The reduction of volatile solvents also includes the use of combinations of solvent- and water-borne coating materials.

2pack epoxy coating materials complying with the VOC guideline

In the following sections, the formulation changes required to reduce the volatile organic solvent content of different resin systems are described, starting with epoxy resins based on standard BPA solid resins.

Epoxy resins for high-solid and solvent-free reaction curing coating materials

For obvious reasons, liquid resins of low viscosity based on BPA were firstly considered for production of low solvent 2pack epoxy systems. Liquid resins showed many disadvantages, such as too high viscosity and significantly longer curing times compared to epoxy solid resins taking four times longer to become tack-free. Additionally, under certain conditions, liquid resins show a tendency towards crystallization, which makes their usage more difficult. Coatings using liquid epoxy resins are hard and brittle and have poor adhesion characteristics because of their higher level of crosslinking [90, 96]. With synthetically combined BPA and BPF liquid resins, the crystallization problem was solved. Combinations with reactive diluents or with plasticized solid resins based on BPA lead to more flexible coatings. 2pack epoxy anti-corrosive coating materials, with optimal characteristic properties, fulfilling the VOC guideline, can be made by combining plasticized epoxy solid resins with liquid resins, reactive diluents and balanced reactive hardeners. The curing of epoxy resin coating materials, especially at low temperatures, can generally be fastened by tertiary amines as catalysts. However, if used in excess, negative coating properties can occur such as reduced corrosion protection, high sensitivity to water, and a high brittleness. These effects can be avoided by using crosslinking accelerators or high speed hardeners [89].

Solvent-free epoxy coating materials can be obtained from liquid resins based on BPA in combination with BPF resins. Aromatic BPA resins are mainly used because of their very good all round durability [97]. Additionally, reactive diluents based on aliphatic glycidyl ethers are used to decrease viscosity and to improve the flexibility of the coatings. These coatings have a higher resistance to solvents and show maximum stability to liquid acids and bases, due to the high crosslinking density, primarily through the usage of BPF resins, which have a higher EP equivalent compared to BPA resins. Compared to solventborne systems, solvent-free epoxy coating materials [97] can be applied in contrast to solventborne systems in very high dry film thicknesses, for example 150 µm and higher in one application. The absence of solvent supports the building of nonporous coatings. This improves the resistance against chemicals and corrosion protection. Additionally, they cure at low temperatures and high relative humidity. Because of the absence of solvents and their high resistance to acids and to bases, solvent-free 2pack epoxy coating systems are used for coating water, beverage and food containers [90].

Epoxy hardeners for high-solid and solvent-free reactive hardening coating materials

The epoxy hardeners define significantly important properties of epoxy coating materials and coatings. They determine the speed of curing and the processing time (potlife) among the characteristics relevant for processing of the coating materials. The properties of epoxy hardeners have a significant influence, amongst others, on the tendency to yellowing, flexibility and the resistance against chemicals [91].

Polyamidoamine (PAA) hardeners: The viscosity of polyamidoamine hardeners, which are proven as standard epoxy resin coating materials, could be reduced by the reduction of the molecular mass and the adaptation of the imidazoline content, but the disadvantage of the incompatibility with liquid resins still exists [89].

Polyamidoamine pre-adduct hardeners: These hardeners are compatible with liquid resins through the pre-addition of lower-molecular weight PAA in benzyl alcohol (BZA). However, they have a relatively high intrinsic viscosity. Benzyl alcohol (BZA) is needed to increase compatibility with liquid resins but is classified as a VOC, therefore, the development of BZA-free hardeners has been necessary and are now available on the market with new, so called, formulated PAA adduct hardeners [98]. Epoxy systems with PAA hardeners cure quickly, show good adhesion and good corrosion protection properties on metals. To obtain fast curing at low temperatures, Mannich bases or Mannich base analogue high speed hardeners are needed in combination with chemically modified PAA with or without benzyl alcohol [89]. Hardeners, for example based on aliphatic polyamines or aromatic polyamine adducts, are used for solvent-free epoxy systems. With these resin hardener combinations, the curing speed and the coating properties like corrosion protection and chemical resistance can be selectively adjusted. Solvent-free 2pack epoxy systems are used for anti-corrosive and chemically resistant coatings in a large range of applications. Examples are: refineries and pipe coatings in the chemical industry, for food containers such as fodder silos and containers for the beverage industry, in shipbuilding, for ballast and chemical tanks, as well as for corrosion protection within steel constructions, for hydraulic engineering including sewer pipes, sewage treatment plants and port facilities as well as desalination plants [97].

To optimize the requirement profile of high-solid 2pack epoxy coating systems concerning processing and coating properties, it is important for paint development to start with the base formulation of the modified solid resins and choose, according to the requirements, other liquid resins as combination partners, e.g. liquid epoxy resins for low solvent coating material [89].

According to the current state of the art, high-solid coating materials for corrosion protection with 65 to 72 vol-% (equivalent up to 85 wt-%) non-volatile components with a volume concentration of pigments (PVC) of 40 % are possible. This corresponds with a VOC content of 235 to 300 g/l when adjusted for processing viscosity, e.g. for airless application [89]. This fulfils the requirements of TL/TP-KOR [86] (see Section 3.2.2.2), e.g. TL sheet 87 and 94 "2pack coating materials based on epoxy resins (lower-molecular weight) and polyurethane , low in solvent content, for maintenance" (TL sheet 94 nm EP/PUR-HS). The solvent contents of these coating systems are for priming coats ≤ 25 wt-%, for intermediate coat(s) ≤ 20 wt-% and for topcoats ≤ 25 wt-% [86], to offer primary protection of steel construction ex factory or for application on the construction area.

High-solid 2pack epoxy coating materials fulfil the requirements for heavy duty corrosion protection and have been used successfully in this field for many years [89, 92]. Table 3.6 shows a guide formulation for a high-solid 2pack epoxy primer for the heavy corrosion protection [99].

The advantages and disadvantages of high-solid 2pack epoxy systems compared to water-dilutable 2pack epoxy systems are shown in Table 3.8.

Epoxy resins and hardeners for water-dilutable 2pack coating materials

The development of water-dilutable 2pack coating materials which started approximately 25 years ago was more challenging at the formulating stage compared to high-solid products because the stability of the emulsified epoxy resin

Table 3.6: Guide formulation for a high-solid 2pack epoxy primer, approx. RAL 1002 [99]

Pos.	Raw material/generic name/trade mark	Parts by weight
Component A (master lacquer)		
01	EP resin, modified based on BPA, semisolid	25.00
02	Thixotropic agent 1	4.50
03	Thixotropic agent 2	1.80
04	Thixotropic agent 3	1.00
05	Pyrogenic silica	1.00
06	Active anti-corrosion pigment: Zinc phosphate	10.00
07	Titanium dioxide, rutile	10.00
08	Filler: talc	10.00
09	Filler: calcit	26.00
10	Iron oxide yellow	2,68
11	Carbon black pigment	0.02
12	Xylene	6.00
13	Isobutanol	2.00
	Sum component A	**100.00**
Component B (hardener)		
14	PAA-adduct-hardener	10.00
	Sum comp. A + B	110.00
01	Epoxy resin EEW: 270-305g/Eq e.g. "Araldite" GZ 290 X 90 (Huntsman)	
02	e.g. "Bentone" 27/10 % in Xylene (Elementis)	
03	e.g. "Bentone" 38/10 % in Xylene (Elementis)	
04	e.g. "Tixcin"/MPA 60 % (Elementis)	
05	e.g. "Aerosil" R972 (Evonik)	
06	e.g. ZPO (Heubach)	
07	e.g. "Hombitan" R611 (Sachtleben)	
08	e.g. AT 1 (Micro Minerals)	
09	e.g. "Millicarb" (Omya)	
10	e.g. "Bayferrox" 930 (Lanxess)	
11	e.g. "Printex" 200 (Orion)	
12	Xylene	
13	Isobutanol	
14	Polyamidoamine adduct: H-equiv.: 115, amin number: 250–290; e.g. "Aradur" 450 (Huntsman)	

To adjust formulation for paint application: thinnable up to 7 wt-% with xylene/n-butanol 4:1.

Characteristics of coating material and cured coating film

- **Typical characteristics of liquid paint** without/with thinner:
  - VOC: 235/300 g/l
  - Solids content: 85/80 wt-%
  - Potlife: > 1,5 h
- **Corrosion protection for a coating structure based on this primer formulation:**
  - Priming coat (PC): 80 µm plus
  - Intermediate coat (IC): 80 µm plus
  - Topcoat (TC) based on 2p-PUR 80 µm
    - 720 h Salt spray ISO 9227,
    - Undercreepage at scribe ≤ 3 mm

and dispersed hardener particles were influenced by the pigments, active anti-corrosion pigments and additives, as well as solvents. Accordingly, the compatibility of these coating components with the binder/hardener system had to be adjusted carefully. Additionally, variations in temperature as well as high shearing forces can adversely influence the stability of the binder emulsion.

Water-dilutable 2pack epoxy paint formulations normally have the hardener component pigmented as this is less sensitive to the above stated influences. Another important factor is the optimal ratio of resin to hardener. The properties concerning good corrosion protection and low water sensitivity are significantly influenced by the degree of crosslinking [100]. Usually optimal results are achieved using 70 to 90 % of the stoichiometric amount of hardener, which has to be optimized on an individual basis by testing [89].

When applying water-dilutable 2pack epoxy systems, the normal indication that the end of the processing time (potlife) is approaching by the increase in the paint viscosity cannot be readily detected which may lead to subsequent coating failures.

PAA hardeners, PAA adduct hardeners, Mannich base and polyamine based hardeners are used for water-dilutable 2pack epoxy systems. PAA hardeners offer good levelling and pigmentation properties. Hydrophilized adduct hardeners using polyamines with low contents of epoxy resins and polyamine hardeners distinguish themselves through a very broad property profile. Hardeners based on Mannich bases are suited for curing at low temperatures because of their higher reactivity. The processing time (potlife) and the minimal temperature for film forming (MFT) of epoxy resin/hardener formulations are influenced by the particle size of the EP resin and hardener as well as by the solvent and catalysts used [101].

The experience gained through the application of the first water-dilutable epoxy hardener systems lead to fundamental technological enhancements particularly connected with the hardener. The development of new emulsion hardeners based on high-molecular aliphatic amines, which can be combined with epoxy solid resin dispersions and EP liquid resin emulsions, provided crucial improvements. The decisive progress in this new hardener technology is the fact, that the hardeners emulsify the epoxy resins during processing very quickly, and they support a very fast curing reaction due to their high molecular weight. The curing time with this new hardener generation is comparable with those of solventborne systems, where the size of the epoxy resin dispersion particles plays a major role, too. The hardener particles have to penetrate into the solid resin. The smaller the diameter of the resin particles, the higher is their overall surface area, which increases the probability of reaction with the hardener particles. With this hardener technology, fast curing with good coating film properties like good corrosion protection, high water resistance, and good wet adhesion with zero VOC values are possible [102].

The so-called second generation of water-dilutable 2pack EP binder systems reached the same level of coating properties, compared to solventborne coatings for stationary coating applications. The current third generation of epoxy binder systems reach the properties of solventborne epoxy systems for non-stationary application conditions, which means field conditions. This enormous increase in quality was possible through the optimization of the epoxy resins, e.g. through an internal flexibilisation of the polymer chain with the extensive abandonment of external hydrophilic components. This minimized the demand of coalescing agents for a proper film building. A further improvement was achieved by the adaptation of the particle size of the EP dispersion. With this, the wet adhesion of water-dilutable 2pack EP systems was increased significantly [103–105] to a degree which is comparable or even better than solvent-based 2pack EP systems.

Besides the choice of EP resin/hardener technology, the extensive optimization of active anti-corrosion pigments and corrosion inhibitors was crucial to the high levels of corrosion protection provided by water-dilutable 2pack EP systems.

Water-dilutable 2pack epoxy primers based on solid resin dispersions generally show good adhesion and good corrosion protection properties on metal surfaces. They generally outperform the properties of comparable solventborne systems concerning wet adhesion and corrosion protection [89, 106].

A new water-dilutable 2pack epoxy resin hardener system called "Easy-Cure-System", referred to in [101, 107] is based on a new polyamine hardener technology which is free of low-molecular amines, does not require special labelling, and is free of organic solvents as well as being stable during freezing and thawing. Good corrosion protection properties of the coating with fast curing or slower curing with higher flexibility of the coating can be obtained by combination of the new hardener technology with different epoxy solid resin dispersions. Because of the high shear stability and good compatibility of solid resin and hardener with pigments, fillers, and different active corrosion protection pigments, the dispersion of pigments can be achieved both within the hardener and within the epoxy solid resin dispersions. The resulting coating materials have very low VOC values of 40 to 150 g/l, due to adaptation with solvents for application with high-rotation bells. Fast curing, no special labelling required and the indication of the end of processing time (potlife) through linear increase of viscosity after mixing of component A and B, makes this an excellent technology. Because of a very good anti-sagging behaviour, dry films of 100 µm can be applied in one step. Very good corrosion protection can be combined with fast curing, e.g. re-coatable after one hour of curing. Table 3.7 shows a guide formulation and variants for faster curing and higher flexibility for a water-dilutable 2pack epoxy resin system using the above mentioned technology and the properties of the coating materials and coatings that can be achieved.

Table 3.7: Guide formulation of a water-dilutable 2pack epoxy resin system [103]

Pos.	Raw material/generic name/trade mark	Parts by weight
Component A (pigmented hardener)		
01	Demineralized water	11.20
02	Wetting and dispersion additive	3.30
03	Defoamer	0.10
04	Talkum (filler)	8.50
05	Titanium dioxide (white pigment)	20.50
06	Iron oxide yellow pigment	0.40
07	Iron oxide black pigment	1.10
08	Barium sulfate (filler)	23.10
09	Active anti-corrosion pigment: zinc/iron phosphate	4.00
10	Corrosion inhibitor	1.35
11	Defoamer	0.05
12	Coalescent agent	0.60
13	Mixture of PUR thickeners	0.60
14	Methoxypropanol	1.00
15	Polyamine adduct hardener, high molecular	24.20
Sum	Component A	100.00
Component B (binder dispersion) "faster curing"		
16b)	BPA solid resin dispersion	48.40
17	Demineralized water	1.60
Sum	Komponente A und B	140.00
Component B (binder dispersion) "higher flexibility"		
16b)	BPA solid resin dispersion	48.40
17	Demineralized water	1.60
Sum	Component A and B	150.00
02	e.g. "Additol" XW 6208 (Allnex)	
03	e.g. "Additol" VXW 6393(Allnex)	
04	e.g. Talkum IT extra (Omya)	
05	e.g. "Kronos" 2190 (KronosTitan)	
06	e.g. "Bayferrox" 3920 (Lanxess)	
07	e.g. "Bayferrox" 306 (Lanxess)	
08	e.g. EWO (Krahn Chemie)	
09	e.g. "Nubirox" 213 (Nubiola)	
10	e.g. "Heucorin" FR (Heubach)	
11	e.g. "Additol" VXW 6393(Allnex)	
12	e.g. "Texanol" (Eastman)	
13	e.g. "Additol" VXW 6388 (Allnex)	
14	e.g. Solvenon PM (BASF)	
15	e.g. "Beckocure" EH 2260w/41WA (Allnex)	
16a)	e.g. "Beckopox" EP 2384w/57WA (Allnex)	
16b)	e.g. "Beckopox" EP 387w/52WA (Allnex)	

Table 3.7 Continue

Typical characteristics of liquid paint and coating, based on this formulation		
	Variant B with "faster curing"	**Variant B with "higher flexibility"**
VOC	approx. 40 g/l	approx. 70 g/l
Crosslinking ratio NH/EP	approx. 0.7	approx. 0.5
Potlife	approx. 1 h	approx. 3 h
Tack-free time	1 h 45 min	3 h
Drying degree (DIN 53150), block resistance after 24 h, 7=best, 1=worst	7	5
Pendelum hardness (König) (DIN EN ISO 1522) after 24 h/7 d	75/110 s	35/105 s
Erichsen cupping test DIN EN ISO 1520 after 7 d drying at RT	1.0 mm	2.5 mm
Corrosion protection of a primer coating dry film, thickness ca. 90 µm on steel, sand blasted - **Undercreepage after 1000 h salt spray test** according DIN EN ISO 9277	2–3 mm	1–2 mm

2pack epoxy formulations that include EP solid resin dispersions exist today that have VOC contents of 40 to 250 g/l [89, 101]. With this, the requirements of the current VOC guideline is fulfilled completely. However, the application of water-dilutable 2pack epoxy resin systems has to be performed at temperatures above 15 °C in order not to fall below the MFT of the coating material. Coating materials for heavy duty corrosion protection, which can be applied outdoors at lower temperatures down to 8 °C, can be combined with epoxy liquid resin emulsions. It should be noted, that curing slows down at lower temperatures but by increasing the temperature, the curing reaction will be accelerated [89]. Very good corrosion protection properties, at least equal to those of solventbased products, can be obtained with water-dilutable 2pack epoxy coating materials based on the state of the art binder and hardener systems on critical substrates like cold-rolled steel, zinc coated steel, and aluminium. For these applications, solid resin dispersions show the best performance [89].

Water-dilutable 2pack epoxy primer coating materials typically contain 60 to 65 wt-% of non-volatile components at the point of use. Wet layer thicknesses of up to 170 µm and dry film thicknesses of 80 µm can be obtained with these materials. The processing time is typically 3 hours at 23 °C and the coatings are recoatable after 16 hours curing at room temperature.

Water-dilutable 2pack epoxy coating systems fulfil the requirements of heavy duty corrosion protection too. This was shown with a formulation based on a modified liquid resin dispersion with a polyamine adduct hardener according to the inspection requests in TL/TP-KOR sheet 87 and sheet 94 in a three layer build containing a 2pack epoxy primer with a primer as an intermediate layer and an 80 µm 2pack-PUR topcoat. Comparative studies with curing at room temperature and at 10 °C and 80 % rh showed no significant differences concerning corrosion protection. Supplemented through tests on Helgoland, which were done over two years in the intertidal zone and splash water zone, the investigations showed that coating systems based on water-dilutable 2pack EP systems are suitable for the highest requirements of corrosion protection as defined in corrosivity category C5 [89].

Table 3.8 shows the state of the art by means of important criteria for 2pack epoxy systems relating to corrosion protection for water-dilutable and high-solid technology according to [89, 101].

Table 3.8: State of the art of water-dilutable and high-solid 2pack epoxy systems relating to corrosion protection by means of important criteria according to [89, 101]

Critera	High-solid 2pack EP	Water-dilutable 2pack EP
VOC values	≤ 230 g/l achievable	VOC = 0 is possible
Availability and variety of binder systems	Limited availability of low viscosity EP resins and hardeners	Solid and liquid BPA based resins and modified types are available. Also different hardener technologies are available
Challenges of coating technology for R&D, paint formulation and application technology	Solventbased, therefore little effort for system changes in application technology	New technology, much effort necessary for R&D, paint formulation and application technology
Toxicology/labeling	Amine of hardeners and reactive thinners are sensitizing and must therefore be labeled	Coating systems without labeling are possible
Processing window	Comparable to standard solvent-based epoxy systems, some limitations by the adjustment of the viscosity	Some limitations: minimum temperature of film building (MFT). Temperature and rel. humidity have to be monitored at processing
Corrosion protection	Very good	Very good
Wet adhesion	Should be tested at galvanized steel and aluminium substrates	Very good on steel, galvanized steel and aluminium substrates
Overpaintablity	Should be tested	Good
Resistance to chemicals	Very good	Chemical performance is slightly under the level of high solid systems

Epoxy coatings – outstanding properties concerning adhesion, wet adhesion and corrosion protection

Epoxy coatings have an extraordinary relevance to corrosion protection because of their outstanding properties. They wet metal surfaces very well and show a tolerance to water and humidity, i.e. they cure on the substrate in the presence of water or humidity [108]. Aromatic, nonpolar structures are a given requirement for hydrophobic behaviour at the interface between the coating and metal.

Another basis for good wet adhesion is given through the high stiffness of the molecules of the cured epoxy system, which is also indicated by relatively high glass transition temperatures (T_g). The bonding groups of the epoxy binder network interact in a synergistic way with the substrate surface. This strong interaction withstands under-creepage of water and electrolyte penetration in an effective way (lateral barrier effect). The humidity resistance and the highly durable adhesive bonds of epoxy coatings on metals, the high mechanical, chemical and thermal stability are the main reasons for the successful use of epoxies for adhesives as well.

The outstanding wet adhesion and the excellent corrosion protection of epoxy coatings are also significant in extremely thin primer layers as found in investigations of extremely thin epoxy resin coatings (0.5 µm) as primer coatings on aluminium and a topcoat of a 2pack-PUR paint (50 µm) [109]. These epoxy coatings with an EP liquid resin, cured with a polyamine show a T_g of 94 °C.

According to [108], the curing of epoxy resin coating materials have a very long sol phase because the gel point is reached only after extensive crosslinking, which has taken place during the curing reaction. Thereby, the main part of the volume shrinkage occurs within the sol phase, so that the remaining shrinkage within the solid phase of the coating is irrelevant. EP coatings are mostly free of internal stress because of this, which promotes a high adhesive strength. The outstanding adhesion of epoxy coatings in contrast to other binder systems is confirmed by other authors, too [110].

3.2.2.1 Heavy duty corrosion protection

Despite the fact that its use is widespread within the coating industry, the term "heavy duty corrosion protection" is not defined clearly. You cannot find it defined explicitly in any corrosion protection standards. The term "heavy duty corrosion protection" means the corrosion protection of steel structures with coating systems according to ISO 12944-5 [111], for applications which demand higher corrosion protection performance than is required from exposure to normal atmospheric conditions, for example, steel construction for hydraulic engineering applications and when exposed to marine conditions. Thus the terms "heavy duty" and "heavy corrosion protection" are synonymous.

Within the standards and guidelines for corrosion protection (see Section 3.2.2.2) a distinction is made between different requirements in atmospheric environments, conditions in marine environments as well as conditions for steel structures embedded in soil. The term "heavy duty corrosion protection" is very common for coatings of outdoors steel constructions and even more frequently in use for coatings under water (steel construction for hydraulic engineering) as well as marine uses. Compared to the corrosion protection required for inland water barrages or water gates, the corrosive stress is much higher for seawater or generally under marine conditions. Offshore constructions are especially exposed to different and changing environmental conditions like immersion in saltwater. Additionally, especially within the splash water area, there are combinations of chemicals (salt, water and oxygen), mechanical and thermal stress [112]. Examples for different corrosive stresses are given in the classification of different corrosion protection requirements for offshore wind energy plants, which are designed for a 20 to 30 years operating life [113]:

- For outdoor areas like the atmosphere over LAT (lowest astronomical tide), corrosivity category C5 M is considered, for water below LAT, category Im 2, and for soil, category Im 3.
- Category C4 relates to inside areas exposed to air,
- category Im 2 for seawater filled excavations, and
- category Im 3 for structures immersed in soil [113].

In order to fulfil these very different requirements, the standards and guidelines for corrosion protection recommend coating systems which include the chemical composition of the individual coating types comprising primer, intermediate and topcoats, in the number and thickness of each layer as well as the total layer thickness of the coating system. These include mostly 2pack epoxy and PUR based primer and intermediate coatings.

2pack epoxy coatings play the major role of providing heavy duty corrosion protection because of their very high performance.

In the above mentioned example of application in offshore wind energy plants the coating structure of choice according to ISO 12944-5:2007 [111] is a system, based on epoxy resins in 2 to 5 layers with a total layer thickness of 500 to 800 µm for underwater areas [114].

2pack epoxy coating systems also play an important role as so-called surface-tolerant coatings (STC). The term STC is often used in connection with maintenance and repair coatings used for heavy duty corrosion protection in shipbuilding and pipeline construction, in offshore steel constructions and other applications, when steel is used in hydraulic engineering constructions [115]. STCs are used as maintenance or repair coatings of objects requiring heavy duty corrosion protection under extreme conditions. These include, when insufficient pretreatment was applied leaving old rusty coatings, when water has condensed on the surfaces to be coated, when there is a high relative humidity and/or low temperature, and high standards of corrosion resistance are required. 2pack epoxy coating materials fulfil these requirements very well because of their specific chemical composition and coating structure, so that they can provide a high barrier effect against water and oxygen and can sustain high thermal and mechanical stresses [116].

Anti-corrosive coatings are a relevant safety factor for the structural safety of steel constructions. In order to ensure that these requirement are met as also identified in ISO 12944 [111] and DIN 55633 [117], the fulfilment of specific tests and testing requirements have to be verified. Additionally, the verification of the structural safety of buildings through layout and design and the quality specifications of the raw materials is an absolute must. These requirements, together with environmental conditions and the associated corrosivity categories and suitable coating systems are described in the comprehensive standard ISO 12944 [68]. Supplementary notes for different applications are stated in ZTV-ING [118] of the German Federal Highway Research Institute (BASt) and Federal Waterways Engineering and Research Institute (BAW) [114]. The essential contents of this and other standards and guidelines for corrosion protection are described in Section 3.2.2.2.

3.2.2.2 Standardized corrosion protection

Corrosion protection of steel has an enormous economic impact, because steel has an outstanding position as a construction material. The corrosion protection through coating systems is one of the most important processes for steel constructions, because organic coatings give the opportunity in a unique way through a versatile choice of product, to get suited corrosion protection, in combination with individual design aspects concerning colouring of the objects to be protected. Because of that, the corrosion protection by coating systems is an important field for standardization.

Due to the excellent corrosion protection properties of coatings on basis of epoxy resin binders, many standards for corrosion protection recommend the use of epoxy resin based coating systems. The most important standards, basic rules and guidelines in connection with corrosion protection by coatings in which epoxy resin systems play a major role, will be described in this chapter. These standards bear references to other standards in the field of corrosion protection, but those would not be explained here.

3.2.2.2.1 Standards for corrosion protection by coatings

ISO 12944 Paints and varnishes – Corrosion protection of steel structures by protective paint systems

This standard describes comprehensively the corrosion protection of steel structures by protective paint systems and is thereby the most important basis or

Table 3.9: Overview of topics of ISO 12944

ISO 12944	Overview of topics of ISO 12944
Part 1	General introduction
Part 2	Classification of environments
Part 3	Design considerations
Part 4	Types of surface and surface preparation
Part 5	Protective paint systems
Part 6	Laboratory performance test methods
Part 7	Execution and supervision of paint work
Part 8	Development of specifications for new work and maintenance

reference for other standards and guidelines concerning corrosion protection. The worldwide valid standard ISO 12944 was adapted to the European EN ISO 12944 and 1998 with the labelling "DIN EN ISO 12944" to the German standard. ISO 12944 consists of the following eight parts, under the general title "Paints and varnishes – Corrosion protection of steel structures by protective paint systems". These eight parts deal with all relevant aspects of corrosion protection of steel structures by protective paint systems (see overview in Table 3.9).

The following parts of the standard will be discussed further, because they contain general information concerning the application area of the standard (part 1), the environmental conditions (part 2), and the requirements for the protective paint systems to be chosen, and recommendations for paint systems (part 5), as well as their tests for approval (part 6).

ISO 12944-1:1998 – General introduction

ISO 12944-1:1998 defines the general application area of all parts of ISO 12944 and contains some basic technical terms as well as a general introduction into the other parts of ISO 12944. Additionally, it contains information concerning health, safety, and environmental protection as well as guidelines how to use ISO 12944 for a certain project.

ISO 12944 deals with corrosion protection by coating systems (organic coatings). Metallic layers are included, too. All aspects, which are relevant for a suitable corrosion protection are covered in the different parts. The essential aim of the standard is described in part 1 as follows:

"In order to ensure effective corrosion protection of steel structures, it is necessary for owners of such structures, planners, consultants, companies carrying out corrosion protection work, inspectors of protective coatings and manufacturers of coating materials, to have at their disposal state-of-the-art information in

concise form on corrosion protection by paint systems. Such information has to be as complete as possible, unambiguous and easily understandable to avoid difficulties and misunderstandings between the parties concerned with the practical implementation of protection work.

This International Standard — ISO 12944 — is intended to give this information in the form of a series of instructions. It is written for those who have some technical knowledge. It is also assumed that the user of ISO 12944 is familiar with other relevant International Standards, in particular those dealing with surface preparation, as well as relevant national regulations.

Although ISO 12944 does not deal with financial and contractual questions, attention is drawn to the fact that, because of the considerable implications of inadequate corrosion protection, non-compliance with requirements and recommendations given in this standard may result in serious financial consequences."

The application area of ISO 12944 is exclusively corrosion protection by paint systems. Thus, other protection functions like resistance against microorganisms, chemicals, mechanical stress and fire are not covered. The field of application is characterized as follows:

- **the type of structure** (made of carbon- or low-alloy steel of not less than 3 mm thickness),
- **the type of surface and surface preparation**,
- **the type of environment**, six corrosivity categories for atmospheric environments and three categories for structures immersed in water or buried in soil,
- **the type of protective paint system**, whereas they dry or cure at ambient conditions. Not covered by ISO 12944 are: powder coating materials, stoving enamels, heat-cured paints, coatings of more than 2 mm dry-film thickness, linings of tanks, products for the chemical treatment of surfaces (e.g. phosphating solutions),
- **the type of work**, and
- **the durability of the protective paint system**.

Durability of the protective paint system

ISO 12944-1 considers three ranges of durability, which is defined as the expected life of a protective paint system to the first major maintenance painting:

- low (L) 2 to 5 years
- medium (M) 5 to 15 years
- high (H) more than 15 years

Durability is a technical consideration, which can help the owner to set up a maintenance program. The durability range is therefore not a "guarantee time" which is the legal subject of clauses in the administrative part of the contract. The guarantee time is usually shorter than the durability range. All paint systems with a range of durability between 5 and 15 years are considered medium. The broad range of durability has to be taken into account by the users and has to be considered during compilation of specifications. It points out that maintenance, which is not related to corrosion protection, can be necessary earlier than provided by the range of durability. E.g. dirt, colour changes, or wear are possible reasons for an earlier maintenance. The possibilities of maintenance or renewal of the paint system have to be considered already at planning and designing, because the reachable duration of protection by paint systems is less than the expected service life of the building. Parts which are exposed to corrosive stress and are not accessible for corrosion protection actions after assembly, have to get such a strong corrosion protection that the structural safety is secured during the service life of the building. The standard commits all persons involved in planning and execution of corrosion protection works to effective actions for health, safety, and environmental protection.

ISO 12944-2:1998 – Classification of environments

ISO 12944-2 describes corrosive stresses, which are caused by the atmosphere and different kinds of water and soil. It defines six corrosivity categories for atmospheric environments (Table 3.10) and three categories for structures immersed in water or buried in soil (Table 3.11), which are described by typical environments in each case. For the influence of different soil types for corrosivity it is referred to DIN EN 12501-1:2003-08 "Protection of metallic materials against corrosion — Corrosion likelihood in soil — Part 1: General". The corrosivity categories are important parameters for the selection of suitable paint systems (ISO 12944-5). It points also to the possible estimation of the corrosivity categories by consideration of the combined effect of the environmental factors like duration of humidification, average concentration of sulfur dioxide, and average area impact of chlorides per year (see ISO 9223).

ISO 12944-5:2007-09 – Protective paint systems

ISO 12944-5 starts with the description of different basic types of paint materials on the basis of their chemical composition and their kind of film building. Additionally, the standard gives examples within comprehensive tables for the selection of paint materials and coating systems used for corrosion protection of steel constructions, which have proven their suitability for the different corrosivity categories with the expected duration of protection. Paint systems as well as metallic layers are used in combination as duplex coatings (duplex systems) for steel constructions. The organic paint systems normally consist of a priming coat, respectively primer, one or more intermediate coats, and a topcoat.

Table 3.10: Corrosivity categories for atmospheric stresses with typical environments according to ISO 12944-2

Corrosivity category	Corrosion load	Examples of typical environments in a temperate climate (informative only) Interior	Exterior
C1	Very low	Heated buildings with clean atmospheres, e.g. offices, shops, schools, hotels	–
C2	Low	Unheated buildings where condensation may occur, e.g. depots, sports halls	Atmospheres with low level of pollution. Mostly rural areas.
C3	Medium	Production rooms with high humidity and some air pollution, e.g. food-processing plants, laundries, breweries, dairies	Urban and industrial atmospheres, moderate sulfur dioxide pollution. Coastal areas with low salinity.
C4	High	Chemical plants, swimming pools, coastal ship- and boatyards	Industrial areas and coastal areas with moderate salinity
C5-I	Very high (industrial)	Buildings or areas with almost permanent condensation and with high pollution	Industrial areas with high humidity and aggressive atmosphere
C5-M	Very high (marine)	Buildings or areas with almost permanent condensation and with high pollution	Coastal and offshore areas with high salinity

NOTES
1 The loss values used for the corrosivity categories are identical to those given in ISO 9223.
2 In coastal areas in hot, humid zones, the mass or thickness losses can exceed the limits of category C5-M. Special precautions should therefore be taken when selecting protective paint systems for structures in such areas.

Table 3.11: Corrosivity categories for water and soil with typical environments according to ISO 12944-2

Corrosivity category	Environment	Examples of environments and structures
Im 1	Fresh water	River installations, hydro-electric power plants
Im 2	Sea or brackish water	Harbor areas with structures like sluice gates, locks, jetties; offshore structures
Im 3	Soil	Buried tanks, steel piles, steel pipes

Important parts and excerpts of this standard are described in the following, at which a special emphasis is taken on the content, which is connected to paint systems on basis of epoxy resins.

Paints for corrosion protection

After the description of paints according to their chemical and physical properties and the kind of constitution, which is important for the application properties, as solventborne, waterborne or solvent-free systems as well as the type of curing/hardening, important binder systems of 1- and 2pack systems are shown.

Annex A of the standard ISO 12944-5 contains examples of practice-proven paint materials for corrosion protection in terms of different environments. Amongst others, acrylic resins, epoxy resin esters, alkyd resins, vinyl chloride polymers, and chlorinated rubber are mentioned as air-drying, respectively oxidative curing paints. As the most important representatives of reactive paints, consisting of master lacquer and hardener, the 2pack epoxy and 2pack-PUR paints are specified.

Typical binder systems as master lacquer component for 2pack epoxy resin paints are: epoxy resins, epoxy vinyl resins, epoxy acrylic resins, as well as epoxy resin combinations (e.g. epoxy hydrocarbon resins). As the most epoxy resin binders and paints are not light stable, it is recommended for outdoor applications to use topcoats based on 2pack PUR systems with aliphatic isocyanate hardeners or topcoats on basis of suited physically drying or waterborne paints. Polyaminoamines (polyamines), polyaminoamides (polyamides) or adducts of these are described as the most common hardener components, at which polyamides are seen as better suited as primers because of their good wettability. Polyamines are mainly recommended for paints with an overall better chemical resistance. This standard talks about products with hardening reactions down to +5 °C. With special products even lower hardening temperatures are possible. 2pack-EP systems are suitable as primers and intermediate coats because of their very good adhesion and wet adhesion.

This standard mentions for 2pack polyurethane paints as typical binders for the master component, polymer binder systems with free hydroxyl groups, which react with suitable isocyanates: Polyester resins, acrylic resins, epoxy resins, polyether resins, fluorinated resins, as well as polyurethane combinations (e.g. polyurethane hydrocarbon resins (PURC)). Aromatic (for primers respectively indoor applications) and aliphatic polyisocyanates are the most commonly used hardener components. Paints cured by aliphatic polyisocyanates (PUR, aliphatic) are used mainly as topcoats because of their very good weather resistance.

Within Section 5 paint systems are described based on environmental conditions and categories of the substrates to be coated. It is differentiated here between initial protection and maintenance, which is stated in annex A with system recommendations. Primers are put into two main categories according to their kind of pigments. These are zinc dust primers (ZN (R)) with an amount of zinc dust within the non-volatile part of the paint of $\geq$ 80 % (wt-%) and other primers which contain zinc phosphate pigments, other corrosion protection pigments or zinc dust at which in the latter case the amount of zinc dust (wt-%) in the non-volatile part of the paint is below 80 %. The paint systems listed in the tables of annex A contain also data, if paint systems with lower content of volatile organic compounds (VOC) like waterborne systems are available for the specified corrosivity categories. Additional information about measures and selection of paint materials in order to achieve further reduction of emissions of VOCs from anti-corrosive coatings are listed in annex D. The suggestions for suitable paint systems include also the required target layer thickness, which is needed to fulfil protection requirements.

Tables of anti-corrosive paints

Section 6 and annex A show instructions for the selection of paint systems as well as examples of paint systems for different environmental conditions and expected durability ranges. Each paint system has to be suited for the highest corrosive stress of the appropriate corrosivity category, which has to be confirmed by the paint manufacturer. The listing of the paint systems in the tables of annex A follows two different aspects:

a) The sorting principle used within Tables A.1, A.7 and A.8 is the reachable protective capacity of the topcoat according to the binder used in it. This choice is best for such cases in which the corrosivity category is not known exactly. This listing contains anti-corrosive systems for multiple corrosivity categories. Table A.1 of this standard is labeled as "summary table", and is shown here within Table 3.12 as a copy from the standard.

b) In contrast to sorting principle in a), the Tables A.2, A.3, A.4, A.5 und A.6 (so-called "individual tables") are based on systems for only one corrosivity category (under the assumption that C5-I and C5-M are one category) and are sorted based on the type of priming coat. The corrosivity category of the environment to which the structure, respectively building to be protected is exposed, is clearly known here by the user.

The following Tables 3.13, respectively 3.14, contain examples of paint systems for an individual or a certain corrosivity category of ISO 12944-5. Additionally, the tables show besides the substrate constitution, how the particular requirements can be fulfilled by a certain number of coating layers and related dry film thicknesses. In addition to the listing of exemplary systems, the tables show in each lower part an overview and a description of binder systems, the number of components, as well as the availability of water-borne coating material alternatives.

In Section 6.2 it is explained that some of the listed systems have a practice-proven durability range of much longer than 15 and partially even more than 25 years. Higher total dry film thicknesses and applying more coating layers will extend

Table 3.12: Table A.1 – Paint systems for low-alloy carbon steel for corrosivity categories C2, C3, C4, C5-I, C5-M from annex A (informative) ISO 12944-5:2007

Substrate: Low-alloy carbon steel Surface preparation: For Sa 2½, from rust grade A, B or C only (see ISO 8501-1)							
System No.	Priming coat(s)				Subsequent coat(s) paint system		
	Binder [d]	Type of primer [a]	No. of coats	NDFT [b] [µm]	Binder	No. of coats	NDFT [b] [µm]
A1.01	AK, AY	Misc.	1–2	100	–	1–2	100
A1.02	EP, PUR, ESI	Zn (R)	1	60 [e]	–	1	60
A1.03	AK	Misc.	1–2	80	AK	2–3	120
A1.04	AK	Misc.	1–2	80	AK	2–4	160
A1.05	AK	Misc.	1–2	80	AK	3–5	200
A1.06	EP	Misc.	1	160	AY	2	200
A1.07	AK, AY, CR [c], PVC	Misc.	1–2	80	AY, CR, PVC	2–4	160
A1.08	EP, PUR, ESI	Zn (R)	1	60 [e]	AY, CR, PVC	2–3	160
A1.09	AK, AY, CR [c], PVC	Misc.	1–2		AY, CR, PVC	3–5	200
A1.10	EP, PUR	Misc.	1–2	120	AY, CR, PVC	3–4	200
A1.11	EP, PUR, ESI	Zn (R)	1	60 [e]	AY, CR, PVC	2–4	200
A1.12	AK, AY, CR [c], PVC	Misc.	1–2	80	AY, CR, PVC	3–5	240
A1.13	EP, PUR, ESI	Zn (R)	1	60 [e]	AY, CR, PVC	3–4	240
A1.14	EP, PUR, ESI	Zn (R)	1	60 [e]	AY, CR, PVC	4–5	320

Table 3.12 Continue

Substrate: Low-alloy carbon steel **Surface preparation: For Sa 2½, from rust grade A, B or C only (see ISO 8501-1)**							
System No.	Priming coat(s)				Subsequent coat(s) paint system		
	Binder d	Type of primer[a]	No. of coats	NDFT[b] [µm]	Binder	No. of coats	NDFT[b] [µm]
A1.15	EP	Misc.	1–2	80	EP, PUR	2–3	120
A1.16	EP	Misc.	1–2	80	EP, PUR	2–4	160
A1.17	EP, PUR, ESI	Zn (R)	1	60[e]	EP, PUR	2–3	160
A1.18	EP	Misc.	1–2	80	EP, PUR	3–5	200
A1.19	EP, PUR, ESI	Zn (R)	1	60[e]	EP, PUR	3–4	200
A1.20	EP, PUR, ESI	Zn (R)	1	60[e]	EP, PUR	3–4	240
A1.21	EP	Misc.	1–2	80	EP, PUR	3–5	280
A1.22	EP, PUR	Misc.	1	150	EP, PUR	2	300
A1.23	EP, PUR, ESI	Zn (R)	1	60[e]	EP, PUR	3–4	320
A1.24	EP, PUR	Misc.	1	80	EP, PUR	3–4	320
A1.25	EP, PUR	Misc.	1	250	EP, PUR	2	500
A1.26	EP, PUR	Misc.	1	400	–	1	400
A1.27	EPC	Misc.	1	100	EPC	3	300
A1.28	EP, PUR	Zn (R)	1	60[e]	EPC	3–4	400

Tabelle 3.12 Continue

Substrate: Low-alloy carbon steel, Surface preparation: For Sa 2½, from rust grade A, B or C only (see ISO 8501-1)																			
System No.	Expected durability (see 5.5 and ISO 12944-1)															Corresponding systems in table				
	C2			C3			C4			C5I			C5M							
	L	M	H	L	M	H	L	M	H	L	M	H	L	M	H	A.2	A.3	A.4	A5 (I)	A5 (M)
A1.01																A2.04				
A1.02																A2.08	A3.10			
A1.03																A2.02	A3.01			
A1.04																A2.03	A3.02			
A1.05																	A3.03	A4.01		
A1.06																		A4.06		
A1.07																A2.03/05	A3.05			
A1.08																	A3.12	A4.10		
A1.09																	A3.04/06	A4.02/04		
A1.10																		A4.06	A5I.01	
A1.11																	A3.13	A4.11		
A1.12																		A4.03/05		
A1.13																		A4.12		
A1.14																			A5I.06	
A1.15																A2.06	A3.07			
A1.16																A2.07	A3.08			
A1.17																	A3.11	A4.13		
A1.18																	A3.09			
A1.19																		A4.14		
A1.20																		A4.15	A5I.04	A5M.05
A1.21																		A4.09		
A1.22																			A5I.03	A5M.0
A1.23																			A5I.05	A5M.06

Tabelle 3.12 Continue

Substrate: Low-alloy carbon steel, Surface preparation: For Sa 2½, from rust grade A, B or C only (see ISO 8501-1)																				
System No.	Expected durability, (see 5.5 and ISO 12944-1)															Corresponding systems in table				
	C2			C3			C4			C5I			C5M							
	L	M	H	L	M	H	L	M	H	L	M	H	L	M	H	A.2	A.3	A.4	A5 (I)	A5 (M)
A1.24																			A5I.02	A5M.02
A1.25																				A5M.04
A1.26																				A5M.03
A1.27																				A5M.08
A1.28																				A5M.07

Binders for priming coat(s)	Paints (liquid)			Binders for topcoat(s)	Paints (liquid)		
	No. of components		Waterborne possible		No. of components		Waterborne possible
	1-pack	2-pack			1-pack	2-pack	
AK = Alkyd	X		X	AK = Alkyd	X		X
CR = Chlorinated rubber	X			CR = Chlorinated rubber	X		
AY = Acrylic	X		X	AY = Acrylic	X		X
PVC = Poly(vinyl chloride)	X			PVC = Poly(vinyl chloride)	X		
EP = Epoxy		X	X	EP = Epoxy		X	X
ESI = Ethyl silicate	X	X	X	PUR = Polyurethane, aliphatic	X	X	X
PUR = Polyurethane, aromatic or aliphatic	X	X	X	EPC = Epoxy combination		X	

[a] Zn (R) = Zinc-rich primer, see 5.2. Misc. = Primers with miscellaneous types of anti-corrosive pigment.
[b] NDFT = Nominal dry film thickness. See 5.4 for further details.
[c] It is recommended that compatibility be checked with the paint manufacturer.
[d] It is recommended for ESI primers that one of the subsequent coats be used as a tie coat.
[e] It is also possible to work with an NDFT from 40 µm to 80 µm provided the zinc-rich primer chosen is suitable for such an NDFT.

Table 3.13: Table A.4 – Paint systems for low-alloy carbon steel for corrosivity category C4 from annex A (informative) ISO 12944-5:2007

Substrate: Low-alloy carbon steel **Surface preparation: For Sa 2½, from rust grade A, B or C only (see ISO 8501-1)**										
System No.	**Priming coat(s)**				**Subsequent coat(s)**	**Paint system**		**Expected durability**		
	Binder	**Type of primer [a]**	**No. of coats**	**NDFT [b] [µm]**	**Binder type**	**No. of coats**	**NDFT [b] [µm]**	**Low**	**Med**	**High**
A4.01	AK	Misc.	1-2	80	AK	3–5	200	■		
A4.02	AK	Misc.	1–2	80	AY, CR, PVC [c]	3–5	200	■		
A4.03	AK	Misc.	1–2	80	AY, CR, PVC [c]	3–5	240	■	■	
A4.04	AY, CR, PVC	Misc.	1–2	80	AY, CR, PVC [c]	3–5	200	■		
A4.05	AY, CR, PVC	Misc.	1–2	80	AY, CR, PVC [c]	3–5	240	■	■	
A4.06	EP	Misc.	1–2	160	AY, CR, PVC [c]	2–3	200	■		
A4.07	EP	Misc.	1–2	160	AY, CR, PVC [c]	2–3	280	■	■	■
A4.08	EP	Misc.	1	80	EP, PUR	2–3	240	■	■	
A4.09	EP	Misc.	1	80	EP, PUR	2–3	280	■	■	■
A4.10	EP, PUR, ESI [d]	Zn (R)	1	60 [e]	AY, CR, PVC [c]	2–3	160	■		
A4.11	EP, PUR, ESI [d]	Zn (R)	1	60 [e]	AY, CR, PVC [c]	2–4	200	■	■	
A4.12	EP, PUR, ESI [d]	Zn (R)	1	60 [e]	AY, CR, PVC [c]	3–4	240	■	■	■
A4.13	EP, PUR, ESI [d]	Zn (R)	1	60 [e]	EP, PUR	2–3	160	■		
A4.14	EP, PUR, ESI [d]	Zn (R)	1	60 [e]	EP, PUR	2–3	200	■	■	
A4.15	EP, PUR, ESI [d]	Zn (R)	1	60 [e]	EP, PUR	3–4	240	■	■	■
A4.16	ESI	Zn (R)	1	60 [e]	–	1	60	■		

Table 3.13 Continue

Binder for priming coat(s)	Type	Waterborne possible	Binder for subsequent coat(s)	Type	Waterborne possible
AK = Alkyd	1-pack	X	AK = Alkyd	1-pack	X
CR = Chlorinated rubber	1-pack		CR = Chlorinated rubber	1-pack	
AY = Acrylic	1-pack	X	AY = Acrylic	1-pack	X
PVC = Poly(vinyl chloride)	1-pack		PVC = Poly(vinyl chloride)	1-pack	
EP = Epoxy	2-pack	X	EP = Epoxy	2-pack	X
ESI = Ethyl silicate	1- or 2-pack	X	PUR = Polyurethane, aliphatic	1- or 2-pack	X
PUR = Polyurethane, aromatic or aliphatic	1- or 2-pack	X			

[a] Zn (R) = Zinc-rich primer, see 5.2. Misc. = Primers with miscellaneous types of anti-corrosive pigments.
[b] NDFT = Nominal dry film thickness. See 5.4 for further details.
[c] It is recommended that compatibility be checked with the paint manufacturer.
[d] It is recommended for ESI primers that one of the subsequent coats be used as a tie coat.
[e] It is also possible to work with an NDFT from 40 µm up to 80 µm provided the zinc-rich primer chosen is suitable for such an NDFT.

Legend for Table 3.14

[a] Zn (R) = Zinc-rich primer, see 5.2. Misc. = Primers with miscellaneous types of anti-corrosive pigments.
[b] NDFT = Nominal dry film thickness. See 5.4 for further details.
[d] It is recommended for ESI primers that one of the subsequent coats be used as a tie coat.
[e] It is also possible to work with an NDFT from 40 µm up to 80 µm provided the zinc-rich primer chosen is suitable for such an NDFT.
[f] Waterborne products are in general not suitable for immersion.

Table 3.14: Table A.6 – Paint systems for low-alloy carbon steel for corrosivity category Im 1, Im 2 and Im 3 from annex A (informative) ISO 12944-5:2007. Legend for Table 3.14 see p. 157.

Substrate: Low-alloy carbon steel
Surface preparation: For Sa 2½, from rust grade A, B or C only (see ISO 8501-1)
Low-durability systems are not recommended and therefore no examples of these are shown.

System No.	Priming coat(s)				Subsequent coat(s)	Paint system		Expected durability		
	Binder	Type of primer [a]	No. of coats	NDFT [b] in [µm]	Binder type	No. of coats	NDFT [b] in [µm]	Low	Med	High
A6.01	EP	Zn (R)	1	60 [e]	EP, PUR	3–5	360			
A6.02	EP	Zn (R)	1	60 [e]	EP, PURC	3–5	540			
A6.03	EP	Misc.	1	80	EP, PUR	2–4	380			
A6.04	EP	Misc.	1	80	EPGF, EP, PUR	3	500			
A6.05	EP	Misc.	1	80	EP	2	330			
A6.06	EP	Misc.	1	800	–	–	800			
A6.07	ESI [d]	Zn (R)	1	60 [e]	EP, EPGF	3	450			
A6.08	EP	Misc.	1	80	EPGF	3	800			
A6.09	EP, PUR	Misc.	–	–	–	1–3	400			
A6.10	EP, PUR	Misc.	–	–	–	–3	600			

Binder for priming coat(s)	Type	Waterborne possible [f]	Binder for subsequent coat(s)	Type	Waterborne possible [f]
EP = Epoxy	2-pack	X	EP = Epoxy	2-pack	X
ESI = Ethyl silicate	1- or 2-pack	X	EPGF = Epoxy glass flake	2-pack	
PURC = Polyurethane combination	2-pack		PURC = Polyurethane combination	2-pack	
PUR = Polyurethane, aromatic or aliphatic	1- or 2-pack	X	PUR = Polyurethane, aromatic or aliphatic	1- or 2-pack	X

the durability range of a paint system. Another way to enhance the durability of a chosen paint system is selecting of a system which is designed for a "higher" corrosivity category than needed, if such a system is used in an environment with lower corrosivity.

Annex D deals with the amount of volatile organic compounds (VOC) within paint materials for corrosion protection and shows possibilities to reduce them by usage of solventborne high-solid paints, solvent-free and water-borne paints, respectively combinations of these paint technologies.

Table 3.15: Test procedure and stress criteria for paint systems applied to steel according to ISO 12944-6

Corrosivity category ISO 12944-2	Durability ranges	Test durations [h] on steel			
		Chemical resistance ISO 2812-1 [a] [h]	Water immersion ISO 2812-2 [h]	Water condensation ISO 6270-1 [h]	Neutral salt spray ISO 9227 [h]
C2	Low (L) Medium (M) High (H)	– – –	– – –	48 48 120	– – –
C3	Low (L) Medium (M) High (H)	– – –	– – –	48 120 240	120 240 480
C4	Low (L) Medium (M) High (H)	– – –	– – –	120 240 480	240 480 720
C5-I	Low (L) Medium (M) High (H)	168 168 168	– – –	240 480 720	480 720 1,440
C5-M	Low (L) Medium (M) High (H)	– – –	– – –	– 720 1,440	480 720 1,440
Im 1	Low (L) Medium (M) High (H)	– – –	– 2,000 3,000	– 720 1,440	
Im 2	Low (L) Medium (M) High (H)	– – –	– 2,000 3,000		 720 1,440
Im 3	Low (L) Medium (M) High (H)	– – –	– 2,000 3,000		 720 1,440

[a] Use method 1 (see 5.6 for the chemicals used). The purpose of the chemical-resistance test is not the assessment of corrosion protection properties but to assess the ability of a system to withstand highly industrial environments. Thus, the test duration remains the same whatever the durability range is. For corrosivity category C5-I, the ISO 2812-1 procedure can be replaced or supplemented by the ISO 3231 test (10 cycles, 240 h for "low" durability; 20 cycles, 480 h for "medium" durability; and 30 cycles, 720 h for "high" durability).

The paint systems shown in Tables A.1 to A.8 of the standard are only examples. Other paint systems with similar protection properties are possible, if it is secured, that the specified durability range can be reached by the paint system. As shown in Tables 3.12 to 3.14, primer paint systems based on epoxy resins are suggested beside PUR and ESI systems for increased requirements concerning corrosion protection (corrosivity category and expected range of durability). At heavy duty corrosion protection within corrosivity category Im 1 to Im 3 (Table 3.14) epoxy resin paints play the dominant role for the suggested priming coat systems.

ISO 12944-6:1998-05: Laboratory performance test methods

This standard specifies laboratory test methods and test conditions for the assessment of paint systems for the corrosion protection of steel structures. Hereby, the resistance against water, moisture, and salt fog is of primary interest as an indication of wet adhesion and barrier properties. The ageing tests and durations specified hereafter have been selected, to ensure, that paint systems do have the characteristics needed for the durability required in the intended application. However, it must be considered, that the active factors in their combination, sequence, and accelerating effect used in artificial ageing tests do not have to match these of natural exposure exactly. Therefore, it is generally recommended, that natural exposure trials of the paint systems should be undertaken. The results of artificial ageing tests shall be used as a resource for an objective comparison of the paint systems to be chosen and not for exact determination of the durability range. They are of special relevance for paint systems without practical experience. It is recommended to use a reference system for the tests too, which is proven in practice over years and whose behaviour during artificial ageing tests is well known. Additional or other tests are recommended in agreement between the interested parties, if necessary.

Tests

Within Section 5 of this standard, methods for accelerated corrosion testing of the paint systems on substrates are described as shown in Table 3.15:

- Water immersion
- Chemical resistance
- Water condensation
- Neutral salt spray
- Water condensation with addition of sulfur dioxide

The duration of these stresses corresponds with the corrosivity category according to ISO 12944-2 and the intended range of durability (low (L), medium (M) and high (H)) according to ISO 12944-1; it increases with growing range of durability as shown in Table 3.15 and 3.16. The current standard calls only for a water

Table 3.16: Stress and testing adhesion of paint systems applied to zinc-coated steel according to ISO 12944-6

Corrosivity category ISO 12944-2	Durability ranges	Water condensation ISO 6270-1
		Test durations [h] on zinc coated steel
C2	Low (L) Medium (M) Low (L)	240 240 240
C3	Low (L) Medium (M) Low (L)	240 240 240
C4	Low (L) Medium (M) Low (L)	240 240 480
C5-I	Low (L) Medium (M) Low (L)	240 480 720
C5-M	Low (L) Medium (M) Low (L)	240 480 720

condensation test for zinc-coated steel according to ISO 6270-1. During revision of this standard in Germany it is in discussion to add salt spray test according to ISO 9227 for testing of zinc-coated steel [119].

Evaluation methods and requirements of paint systems after ensued stress tests are shown in Table 3.17. The structured description of the testing results in annex A of the standard allows a consistent documentation and classification of the paint systems according to their performance in corrosion protection.

The series of standards of ISO 12944 is well proven in praxis and is a solid base for other standards, technical guidelines, regulations, specifications and conditions of contract concerning corrosion protection of steel structures. Because of its comprehensive description of the different aspects and frame conditions and its classification, it is cited in specialist publications, too [121]. Part 5 and 6 of this standard are revised at the moment in Germany by DIN [119]. A compilation of this and other standards for corrosion protection can be found within DIN-paperback 286 [122].

DIN 55634:2010-04: Paints, varnishes and coatings – Corrosion protection of supporting thin-walled building components made of steel

This standard covers corrosion protection of supporting thin-walled building components made from non-alloy or low-alloy steel with a nominal sheet thickness of up to 3 mm and which are exposed to atmospheric corrosive stress. It contains coil and piecework coatings, which were made in the factory or at the building site.

Table 3.17: Evaluation methods and requirements of paint systems according to ISO 12944-6 [120]

Assessment methods		Requirements
Assessment before artificial ageing		
ISO 2409	Cross-cut classification	0 or 1 for DFT < 250 µm; if DFT > 250 µm the following test shall be used instead of ISO 2409
ISO 4624	Adhesion	No adhesion break to the substrate (A/B) allowed (unless pull-off values are 5 MPa or more)
Assessment after artificial ageing for the specified time		
ISO 4628-2	Blistering	0 (S0) (assessment immediately)
ISO 4628-3	Rusting	Ri0 (assessment immediately)
ISO 4628-4	Cracking	0 (S0) (assessment immediately)
ISO 4628-5	Flaking	0 (S0) (assessment immediately)
Complementary assessment methods after 24 h reconditioning at normal climate acc. 5.4		
ISO 2409	Cross-cut classification*	0 or 1 for DFT < 250 µm; if DFT > 250 µm the following test shall be used instead of ISO 2409
ISO 4624	Adhesion	No adhesion break to the substrate (A/B) allowed (unless pull-off values are 5 MPa or more)

* Film thickness > 250 µm: cross-cut according to DIN EN ISO 16276

Corrosion protection systems for these applications are metallic coatings, paints and metallic coatings with paints (duplex systems). In contrast to ISO 12944, powder coatings are covered in this standard, too. The corrosivity categories from ISO 12944-2 are applied here, too. Additionally, and only for coil coatings, this standard applies according to DIN EN 10169-2 for a corrosion resistance category. The declaration of the corrosion protection class is used only for the assignment of previous building-authority requirements to the new European classification system, consisting of corrosivity category and range of durability.

This standard can also be used for nonstructural thin-walled building components. Both, powder coatings according to DIN 55633:2009-04 and liquid coatings according to ISO 12944-5, are used for piecework coating. A large share of the coating materials recommended in this standard is based on epoxy resins.

The coating systems used for coil coatings are usually named by the binders of the topcoat (see DIN EN 10169-1:2004-04, annex A). Further details of coil coatings are described within Section 3.5.

DIN 55633:2009-04: Paints and varnishes – Corrosion protection of steel structures by powder coating systems – Assessment of powder coating systems and execution of coating

This standard covers the use of powder coating materials for structural steel work. It supplements the series of standards ISO 12944, which covers liquid coating materials only.

DIN 55633 covers steel constructions, which parts are made of non-alloy or low-alloy steel (e.g. according to DIN EN 10025-1:2005-02) with a minimum thickness of 3 mm, which are designed according to an assessment of load-bearing stability and are unpainted or zinc coated (hot-dip-galvanized). The following basic types of powder coating materials are recommended for corrosion protection of steel constructions. The film formation or curing results by thermal crosslinking.

Typical binders for powder coatings

Epoxy resins (EP), epoxy/polyester resins (EP/SP or hybrid binders), polyester resins (SP) as well as isocyanate curing hydroxy-functional polyester resins (PUR) are given as typical binders for thermal curing at 150 °C to 220 °C. Based on the properties of the particular binder system, epoxy powder coating materials are used mainly as priming coats, and polyester respectively PUR powder coating materials are mainly used for topcoats. For execution of laboratory tests for assessment of the coating systems, ISO 12944-6 is referred. Annex A shows examples of powder coating materials. An excerpt of these is specified in Table 3.18. The examples of powder coating systems in this standard are more or less identical to those of DIN 55634.

ISO 20340:2009-04: Paints and varnishes – Performance requirements for protective paint systems for offshore and related structures

This international standard describes the performance requirements, test procedures and assessment of testing results of protective paint systems for offshore and related structures. It covers objects and plants, which are in marine atmosphere (according to atmospheric corrosivity categories C5-M) and immersed in seawater or brackish water within the range between low and high water level (respectively immersion category Im 2) from ISO 12944-2. Additionally it is focused on high-durable paint systems with the aim to reduce costs for maintenance and specifies additional testing requirements, which are over and above those specified for corrosivity category C5-M high durability in ISO 12944-6. This standard includes:

- Test procedures to determine the composition of the separate components of the protective paint system
- Laboratory test procedures for assessment of the likely durability of the protective paint system
- Criteria for assessment of test results concerning resistance

Table 3.18: Examples for powder coating systems from DIN 55633:2009-04

No.	Surface preparation or pretreatment[a]	Primer coating(s)			Topcoat(s)			Coating system	
		Binder type	No. of coats	NDFT [µm]	Binder type	No. of coats	NDFT [µm]	No. of coats	NDFT total [µm]
P1.1	Sa 2 ½	EP	1	80	–	–	–	1	80
P1.2	Sa 2 ½	–	–	–	EP/SP, SP, PUR	1	80	1	80
P1.3	Sa 2 ½		–	–		2	60	2	120
P1.4	Sa 2 ½	EP	1	60		1	60	2	120
P1.5	Sa 2 ½		1	80		1	80	2	160
P1.6	Sa 2 ½		1 – 2	100		1	80	2 – 3	180

[a] Alternative, similarly suitable preparation and pre-treatment processes may be agreed.

EP = Epoxy resin
EP/SP = Epoxy/polyester hybrid resins
SP = Polyester resin
PUR = Polyurethane (isocyanate hardening OH functional polyester resin)

L = low
M = medium
H = high

No.	Expected protective life (ISO 12944-1)														
	Corrosivity category														
	C2			C3			C4			C5I			C5M		
	L	M	H	L	M	H	L	M	H	L	M	H	L	M	H
P1.1	■	■		■											
P1.2	■	■													
P1.3	■	■	■	■	■										
P1.4	■	■	■	■	■	■	■								
P1.5	■	■	■	■	■	■	■	■	■	■			■		
P1.6	■	■	■	■	■	■	■	■	■	■	■		■	■	

Table 3.19: Minimum requirements for protective paint systems according to ISO 20340:2009-04

Substrate	**Blast-cleaned carbon steel: Sa 2½ or Sa 3; surface profile: medium (G)**							**Hot-dip-galvanized steel or steel with Zn-based metallizing** [a]
Corrosivity category of environment	C5-M		Splash and tidal zones C5-M and Im 2			Im 2		C5-M
First coat	Zn (R)[b]	Other primers[c]	Zn (R)[b,d]	Other primers[c]		Other primers		
NDFT [µm]	≥ 40	≥ 60	≥ 40	≥ 60	≥ 200	–	≥ 150	
Minimum number of coats [e]	3	3	3	3	2	1	2	2
NDFT of paint system [µm]	≥ 280	≥ 350	≥ 450	≥ 450	≥ 600	≥ 800	≥ 350	≥ 200
Minimum pull-off test value (before ageing) determined in accordance with ISO 4624 [MPa]	3	4	3	4	4	8	4	3

[a] The thickness of the metallic coating shall be in accordance with ISO 1461 (hot-dip galvanized) or ISO 2063 (metallized steel) and the coating shall be prepared as specified in ISO 12944 4:1998, Clause 12 (hot-dip galvanized) or Clause 13 (metallized steel). Overcoating of thermally sprayed aluminium (TSA) is not recommended due to the risk of the overcoat flaking and corrosion of the TSA occurring. For TSA, a sealer coat only is recommended.

[b] Zn (R) = Zinc-rich primer as defined in ISO 12944 5:2007, Subclause 5.2 (minimum 80 % by mass of zinc dust in the non-volatile part of the paint). The zinc dust pigment shall conform to ISO 3549.

[c] The use of primers other than Zn (R) is mainly applicable to repair and maintenance. For new constructions, the use of other primers should be restricted to areas subjected to special stresses (as defined in ISO 12944 2:1998, Annex B, Clause B.2), where the need for a coating system with higher mechanical strength or higher chemical resistance outweighs the better rust creep protection offered by Zn (R) primers. Examples of areas subject to special stresses are helicopter decks, splash and tidal zones, walkways, escape routes, material lay-down areas and mud zones.

[d] This coating system with an organic Zn (R) primer can also be used for Im2 service if a Zn (R) primer is desired. In this case, the NDFT of the complete system can be reduced to ≥ 350.

[e] The number of coats does not include a tie coat, which might be needed when a Zn (R) silicate primer is used, for instance.

In many contents this standard refers to ISO 12944, like e.g. in Section 4 "Field of applications" to environmental conditions (ISO 12944-2) and for general selection of the kind of coating material (Section 4.5) to ISO 12944-5. It demands detailed information of the coating materials, like amongst others for labelling and for important product-specific and technical data of the coating materials, the generic name of the coating materials and important components. Each paint and varnish within a coating system has to perform both, a "fingerprint-check" and a "routine batch-check" at first and subsequent deliveries. Annex B contains the minimal amount of the fingerprint-check, amongst others, as an identity check an infrared spectrum, the non-volatile mass percentage, kind and percentage of pigments and functional groups of the binders: Epoxy, OH, acid and isocyanate groups.

Section 6 describes corrosion protection systems together with environmental conditions, substrate and its pretreatment as well as product information for each single layer in the coating structure. Within Section 6.2 minimal requirements for coating systems are defined, which are summed up in Table 3.19. The number of layers and the total layer thickness are primarily essential for the durability range.

The qualification tests of coating materials (Section 8) are summed up in Table 3.20. Additional optional tests, like chemical resistance, resistance against abrasion or cracking can be performed, if agreed between the contract partners.

Table 3.20: Qualification tests according to ISO 20340:2009(E)

Test	Scribe line	Environment of corrosivity category C5-M	Environment of combined corrosivity category C5-M and Im 2 (splash and tidal zones)	Environment of corrosivity category Im 2
Ageing resistance (see Annex A)	Yes (see 8.1.8)	4,200 h	4,200 h	–
Cathodic disbonding (ISO 15711:2003, method A, unless otherwise agreed)	No (artificial holiday used instead–see Table 5)	–	4,200 h	4,200 h
Sea water immersion (ISO 2812-2)	Yes (see 8.1.8)	–	4,200 h	4,200 h

The evaluation and the requirements of the qualification tests are performed according to ISO 12944-6 (see Table 3.17).

To determine the ageing resistance (annex A) the corrosion protection systems are exposed to the following cyclic corrosive stress, where one test cycle has a duration of 168 h (one week):

- 72 h exposure to UV-rays and condensation according to ISO 11507:2007 under the following conditions:
  - Method A from ISO 11507: Alternating periods of 4 h UV-stress at 60°C and 4 h stress through condensation of water at 50 °C
  - Type II UV-lamps (UVA-340) according to ISO 11507-Section 5.1.2
- 72 h exposure to salt spray test according to ISO 9227
- 24 h exposure to low temperature at -20 °C

3.2.2.2.2 Standards and guidelines for anti-corrosive coatings

The corrosion protection of steel and also economically important aluminium materials is the subject of standards, guidelines, technical test specifications, and delivery conditions of associations and institutions. These writings, which are mainly published as brochures, are intended to help those which are working with corrosion protection within planning, construction, tendering, execution, and, to get basic knowledge about corrosion protection, the most important influencing variables, and requirements as well as becoming familiar with important guidelines of standards and material respectively substance properties. Furthermore they give concrete assistance to tender corrosion protection works, about delivery conditions, product specifications, test conditions and scopes, as well as for execution and monitoring of corrosion protection works. Thereby they provide an indispensable contribution for standardization and maintenance of a durable and high quality corrosion protection.

Additional technical terms of contract and guidelines for engineering structures ZTV-ING from the Federal Highway Research Institute (BASt)

Important standards for corrosion protection are "additional technical terms and guidelines for engineering structures (ZTV-ING)" as well as the TL/TP-KOR listed below which are worked out by the Federal Highway Research Institute (BASt) and are published by the German Ministry of Transport and Digital Infrastructure [123].

In the following, the standards and policies of BASt [125] are cited from BASt homepage: "The standards and policies for the planning of corrosion protection and execution of corrosion protection measures is based on an extensive collection of provisions, consisting of ISO and DIN standards and supplementary sets of regulations. The eight-part ISO 12944 standard lays down basic principles regarding corrosion protection in layers.

The supplementary sets of regulations on the subject of corrosion protection for the area of federal transport routes are:

- Additional Technical Terms of Contract and Guidelines for Civil Engineering Works (ZTV-ING), Part 4 Steel Structures, Composite Steel Structures, Section 3 Corrosion Protection of Steel Structures, as of 12/2012 (ZTV-ING 4-3)[123].
- Technical Terms of Delivery and Technical Test Regulations for Coating Materials for the Corrosion Protection of Steel Structures, 2002 edition (TL/TP-KOR-Steel Structures)[86].
- Guidelines for Corrosion Protection of Steel Ropes and Cables in Bridge Building, 1983 edition, (RKS-steel ropes) henceforth ZTV-ING Part 4-4 Steel Ropes and Cables, ZTV-ING Part 4-5 Corrosion Protection of Steel Ropes and Cables and TL/TP-KOR-Steel Ropes and Cables.
- Guidelines for Maintaining Corrosion Protection of Steel Structures (RI-ERH-KOR).

The BASt undertakes the updating of these sets of regulations based on the continuing development of the state of the art in corrosion protection, professional expertise and its own investigations.

Corrosion protection coatings:

The choice of corrosion protection system depends on a number of factors, including corrosive stress (exempli gratia effect of road salting, impact of chippings, dampness from structural components in contact with the ground), the planned service life of the component and associated considerations on the corrosion protection maintenance schedules (duration of protection), size of the structural component, approachability to the surfaces or accessibility.

ZTV-ING 4-3 contains examples of suitable corrosion protection systems for structures and structural components, which take these factors into account. The composition and characteristics of coating materials are described in the TL/TP-KOR Steel Structure guidelines in so-called TL sheets, from which these corrosion protection systems are produced[86, 126, 127]. At present, there exist a total of 14 TL sheets. Coating materials complying with the requirements laid down in the TL/TP-KOR Steel Structures guidelines are listed in a "Compilation of certified coating materials complying with the TL/TP-KOR Steel Structures guidelines" maintained by the BASt[128]. The inclusion in this compilation ensures the presentation of appropriate testing certificates, as well as the proof of external monitoring. The coating materials taken from this compilation are usually used on structures and structural components along German federal transport routes.

Two-component materials:

Thus, for example, 2pack materials (Sheet 87, Sheet 94, Sheet 97) with an epoxide resin priming coat and zinc-dust-pigmented primer, one or two micaceous iron oxide intermediate coats, and a final light-resistant 2pack polyurethane coat are mostly used as corrosion protection on large objects with a large surface area requiring a major overall effort. The coatings used for instance on bridges in accordance with Sheet 87 have an excellent track record extending over 30 years. When properly executed, such corrosion protection coatings have a minimum protection period of 25 years and are regarded as the benchmark for comparable investigations of new coating systems. The coating materials in accordance with Sheet 94 represent a further development of Sheet 87 in the direction of less use of solvents. Sheet 97 has been especially developed for use in low temperatures and faster treatment in the factory than Sheet 87.

These epoxide resin and polyurethane coating materials are all extremely resistant and therefore also suitable for use on the undersides of orthotropic road bridge decks, which can be stressed with temperatures up to 150 °C on installing the hot mastic asphalt layers."

In Section 3.2.2, Table 3.5, the requirements for the composition, the properties of the liquid paint and the coating structure of the coating system of TL sheet 87 "EP/PUR" "Coating materials on basis of epoxy or polyurethane (high-solid coating materials)" are described.

The TL sheets are revised according to the state of the art. Currently TL sheet 87 is revised especially with regard to intercoat adhesion [129].

The ZTV-ING is not valid for hydraulic structures. The additional technical terms – hydraulic engineering (ZTV-W) for corrosion protection within hydraulic steelwork (performance range 218) [124] is treated in the following.

Rules and standards of Federal Waterways Engineering and Research Institute (BAW)

The Central Waterways Engineering Library (VZB) as the scientific library of the Federal Waterways Engineering and Research Institute documents and issues literature, information also from the Federal Waterways and Shipping Administration (WSV). From VZB, among others, following information are available:

Guidelines, rules, standards, technical recommendations, and codes of practice ("Merkblätter" data sheet) of the BAW, for example:

- "Technisches Regelwerk Wasserstraßen (TR-W)" [BAW-Website]
- "Standardleistungskatalog für den Wasserbau (STLK-W)" [BAW-Website]
- BAW "Guideline for the Testing of Coating Systems for the Corrosion Protection of Hydraulic Steel Structures (RPB)" [130]

Brochures about corrosion protection of steel structures by protective paint systems

These brochures, which are published from the German Society for coatings and printing inks industry (VdL) together with the Federal Society for Corrosion protection e.V. (BVK) [120], from German Steel-Information Center [131] and the guideline corrosion protection of steel structures under atmospheric environments by protective coatings published by Deutscher Stahlbauverband (DSTV) and Institut für Stahlbau, Leipzig [132] describe very clearly and concisely all aspects of corrosion protection of steel structures by coating systems according to ISO 12944.

A good overview of the corrosion protection of steel structures with a special focus on metallic coatings and subsequent application of coating systems is found in the brochure, the DSTV, the BVK, VdL and the Industrieverband Feuerverzinken have jointly issued [133].

Quality and inspection regulations from the Association for High Quality Corrosion Protection of Valves and Fittings with Powder Coating (GSK)

GSK as a trade association in Germany acts as the global network for companies that coat valves and fittings with epoxy powder coatings. The GSK provides advice on epoxy powder coatings to help engineers, developers, construction companies, manufacturers, local authorities, and operators of water and gas mains all over the world. GSK is also pursuing the objective to ensure and improve the quality of the coatings of fittings in pipe networks of water and gas with GSK's Quality and Inspection Regulations. For instance, the RAL-GZ 662 from GSK guarantees a tested, faultless coating quality and consequently a future reliable solution for the mains supply grid. The RAL-GZ 662 is only specified for manufacturers who wish to meet the highest requirements of a comprehensive quality assurance [134].

The GSB International Quality Regulations

The international quality regulations GSB AL 631 are applicable to the piecework coating of aluminium building components and its alloys used in structural engineering. It includes part-finished products, part-finished and finished piecework as well as prefabricated construction parts of uninstalled building components. Coil coating and on the spot technical coating methods are excluded.

GSB ST 663 applies to the industrial piecework coating of steel components, where coatings applied manually on site are excluded. Coatings governed by these quality regulations are applied to non-galvanized steel surfaces, strip alloy galvanized steel surfaces (DIN EN 10326, and DIN EN 10327), and hot dip galvanized steel surfaces (DIN EN ISO 1461) and covers finished piecework as well as prepared structures.

The quality guidelines for aluminium AL QR 631 and for steel and galvanized steel QR ST 663 [136] include in addition to the requirements for the coatings also the principal processes and specifications for job coater plants, galvanizing plants, and pretreatment and coating material manufacturers. Furthermore, they shall determine the conditions for obtaining of quality labels. In Section 7 examples of powder coatings and liquid coating systems for galvanized steel (duplex systems) are given. Here, epoxy priming coats are recommended with suitable topcoats in both, liquid as well as powder coatings, at high expected durability of protection.

The Quality Regulation IQC 654 applies to the industrial coating with powder coatings of piecework made from steel, galvanized steel, or aluminium and its alloys piecework [136].

3.3 Applied flooring technology

Rob Rasing

3.3.1 Concrete

Concrete is a composite material of gravel, sand, water and cement. The simple process of mixing and curing of cement was already applied by the ancient Greeks and Romans [137]. The Pantheon in Rome, for example, is one of the largest, unreinforced concrete structures in the world today [138]. After the fall of the Roman Empire, most know-how was lost and it was only until the 18th century before concrete became popular again as a building material [139].

Today, steel-reinforced concrete slab is the most used floor substrate for use in residential, commercial and industrial sectors [140].

Though concrete as a construction material is considered inert during most of its service life, as a substrate it is never static. The characteristics of a concrete slab are impacted by many factors such as the use of a waterproofing membrane, the concrete pour and the preparation of the concrete surface after cure [141, 142]. The concrete pour balances a number of variables, which impact concrete's porosity and mechanical strength in particular. When properly designed, executed and fully cured, reinforced concrete is a versatile construction material that provides high mechanical strength and modulus over a long time span [143]. However, during its service life, concrete's strengths are countered by a variety of exposure mechanisms. The most common ones include: chemical exposure, carbonation and chloride-induced deterioration; freeze-thaw deterioration; mechanical wearing; thermal exposure and cracking. The chemical, physical and thermal degradation mechanisms, if not dealt with, can ultimately compromise a structure's integrity.

Architects, specifiers and end-users can select from a range of technologies and finishes to protect concrete slabs. This chapter outlines amine-cured epoxy resin as a polymer technology for use in industrial flooring application.

3.3.2 Application of epoxy thermosets for ambient cure condition

The cure of epoxy resins is a complex process [144] and it will help the reader to understand the basic principles in more detail. A good outline of cure fundamentals of epoxy thermosets is summarized by a time-temperature-transformation (TTT) diagram illustrated in Figure 3.5 [145, 146]. The diagram describes changes in the states of a thermoset curing material as a function of time and temperature. The S-shaped vitrification contour and the gelation contour dominate the cure pathway. The glass transition temperature, T_g, can be utilized as a practical parameter for following the cure process of reactive thermosetting systems. Three critical temperature points mark the y-axis: T_{g0}, the glass transition temperature of the starting materials; gel T_g, the temperature where gelation and vitrification coincide; and $T_{g\infty}$, the glass transition temperature of the fully cured system.

Consider a flooring application where the cure temperature, T_{cure} is constant and approximates ambient conditions. At the start, the condition $T_{cure} \gg T_g$ is met. Components are mobile in a liquid state and reaction rates are dictated by kinetics. The material's T_g readily increases to the gelation point, gel T_g. As chemical conversion increases towards $T_{g\infty}$, mobility of components is reduced and reactions become diffusion-controlled. In the later stages of cure, in particular after vitrification, reaction rates become extremely low. The material

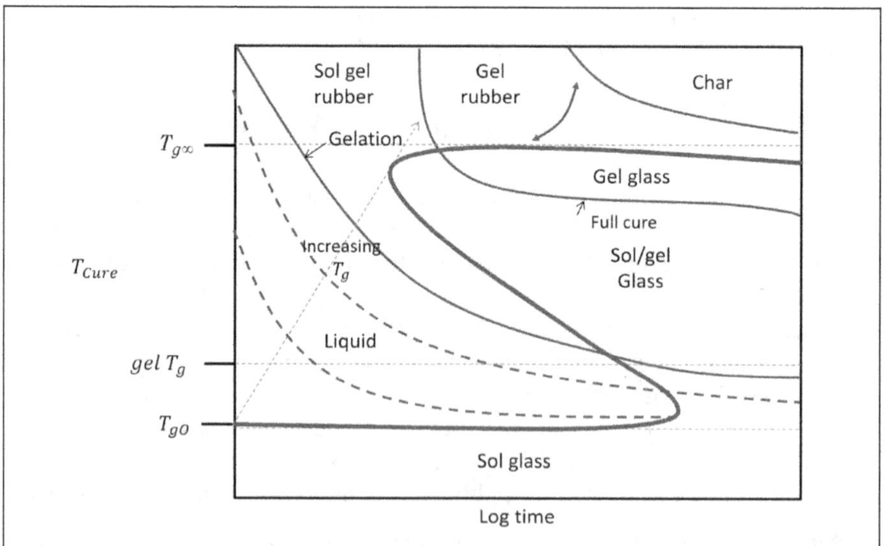

Figure 3.5: A generalized TTT diagram for an isothermal curing thermosetting system. The dashed lines represent iso-T_g contours [145].

appears to be in a 'frozen' state, often referred to as B-stage. Once entered a B-stage, the material will no longer be able to reach full cure, unless conditions are radically adjusted (e.g. provide a heat post-cure).

Flooring application is typically based on a liquid epoxy resin with an EEW 182-192 (LER, Table 2.10). Use of a reactive diluent at 10 to 15 weight-% of the total composition provides the necessary handling properties at ambient and lower temperatures. In general, C_{12} to C_{14} alkyl glycidylether are a popular choice of diluent for balancing cost and performance. Furthermore, the epoxy resin is preferably cured with an amine to form a high molecular weight polymer. Often, the amine needs advancement through chemical modification and formulation to enable a thermoset material with high chemical conversion that delivers the required performance. In this respect, a large temperature delta between the temperature of the floor environment and the polymer T_g is fundamental. At the same time, vitrification (B-stage) of the cured material has to be avoided. Though B-staged material looks identical to fully cured material, it provides inferior mechanical and chemical integrity among other properties.

Figure 3.6 illustrates the necessity for chemical modification of amines for use in ambient cure thermosets [147]. The amines (tetraethylene triamine, TETA; polyoxypropylene diamine MW≈230, D230; and 4,4'-methylenebis(cyclohexylam ine), PACM) are mixed at stoichiometric amounts with unmodified epoxy resin and cured at 23 and 10 °C. At ambient temperature cure, the cured material develops an actual T_g around 50 °C, which lies about 20 to 30 °C higher than the applied cure condition. The full cure state ($T_{g\infty}$) is only achieved after applying the necessary heat cure. The gap between actual T_g and $T_{g\infty}$ (ΔT_g) represents residual cure that has not materialized at the ambient cure conditions and is a first indication of a B-staged material.

Chemical engineering of amines is the methodology to minimise ΔT_g and develop products that meet handling and performance criteria when cured at ambient conditions. The formulated amines are referred to as amine curing agents and are used for high performance industrial flooring application. In general, curing agents are derived from (cyclo)aliphatic amines, polyamides and amidoamines, and Mannich bases. In addition, waterbased amine curing agents have proven valuable, both from a handling and a performance point of view [148].

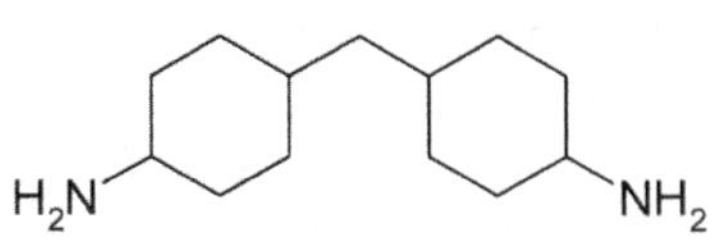

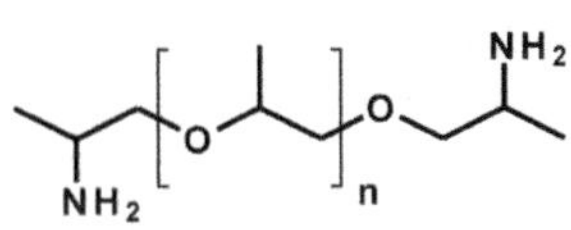

Equation 3.15: 4,4'-methylenebis(cyclohexylamine) (PACM), AHEW 52.5 (left); and a polyoxypropylene diamine (D230), AHEW 58 (right).

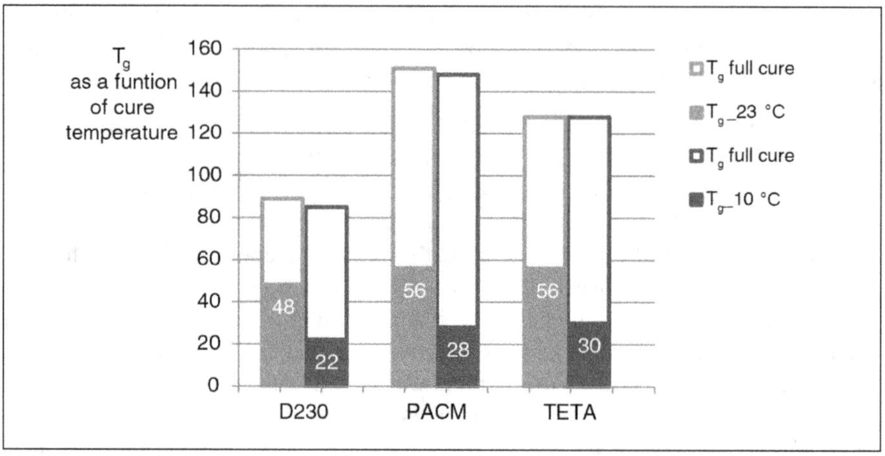

Figure 3.6: T_g of amine-cured epoxy resin (DGEBA) as a function of cure temperatures (23 and 10 °C)

3.3.3 Floor design and installation

The value chain of industrial flooring application is a useful approach to understand relations between the various parties involved (Figure 3.7). The actual floor system application is the centre point where requirements of applicator (installer), architect (specifier) and end-user meet. The requirements of each party are summarized in Table 3.21.

End-user and architect

The end-user, often the object owner, has a demand for an industrial floor system. This can be a new floor system to replace an existing floor or a new building with installation of a floor on recently-poured concrete. Particularly concerning replacement jobs, the end-user often takes the role of specifier with a primary focus on floor performance during the service life and minimising downtime. The latter is also referred to as 'back in service time'. Table 3.22 shows an overview of performance requirements and attributes, which will be further discussed in Chapter 3.3.4. Aesthetics often play a secondary role in refurbishment jobs and depend on the type of industrial floor application as outlined in Chapter 3.3.5.

In case of new buildings and structures with highly specific performance demands, a dedicated specifier or architect is often involved with floor design. With new buildings, downtime is less of a concern since there is no disruption of running business activities. Architects often require elements including floorings to fit within their design. Frequently, new trending features and aesthetics come with higher priority compared to renovation jobs.

Applicator (installer)

The applicator is responsible for installing the industrial floor with respect to the specifications. Logically, his interest lies in the direct handling and application features of the materials worked with. The handling properties are a compilation of a group of product properties, namely:

- the personal protective equipment (PPE) required working with the materials;
- the preparation and mixing of the components;
- the mixed viscosity of the material;
- the application of the material with roller, squeegee or serrated trowel;
- the working time of the material (referred to as potlife); and
- the amount of solvents released during application as specified by the volatile organic content (VOC).

The applicator works under incumbent conditions of temperature and humidity. In practise this means that materials are applied between 8 and 35 °C and under a range of relative humidities. The scenario of low temperature and high humidity, generally described as 'adverse conditions', is the toughest for delivering good performance. Low temperatures extend the potlife and cure times while they result in higher mixed material viscosity. High humidity can cause problems with appearance and adhesion as discussed in the next chapter.

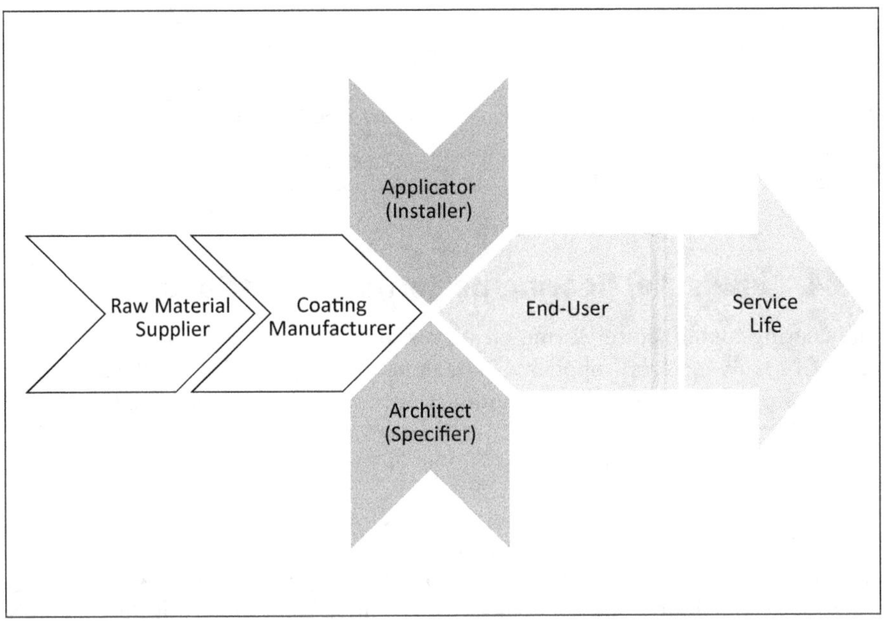

Figure 3.7: Value chain of applied flooring technology *Source: Air Products and Chemicals, Inc.*

Table 3.21: Prototype targets for industrial flooring across the value chain.

	Floor design	Floor installation	Service life
Handling properties		•	
Cure under adverse condition		•	
Back in service time		•	•
Appearance (aesthetics)	•	•	•
Performance over service life	•		•

Table 3.22: Performance testing standards and references

	Reference or Industry Standard
Drying, cure and hardness build	ASTM D1640 (Thumb-twist method) ISO 1522 (Pendulum hardness) ISO 868 (Shore A/D hardness)
Carbamation and water spotting	ISO 2812 (Early water resistance)
Adhesion strength	ISO 2409 (Cross-hatch) ISO 4626 (Pull-off testing)
Mechanical resistance	ISO 178 (Flexural mode) ISO 527 (Tensile mode) ISO 604 (Compressive mode)
Conductivity	DIN-EN 61340-4 and -5
Wear resistance	ISO 7784 (Abrasion resistance) EN ISO 1518 (Scratch resistance)
Chemical resistance	ASTM D543 (Immersion method)
Temperature resistance by heat deflection	ASTM D648
UV resistance	EN-ISO 11507 (Exposure of coatings) EN-ISO 7724-2 (Colorimetry)
Emissions and indoor air quality	EN-ISO 16000

3.3.4 Industrial flooring performance attributes

The coating manufacturer formulates floor materials using the various components from raw material suppliers. Though not directly involved in the actual installation of the floor system, his product needs to meet all requirements as set forth in Table 3.21. The following concise overview expands on the performance attributes of Table 3.22.

Adhesion

Adhesion is the most fundamental property of the binder system when coating a substrate. Inherently, loss of coating adhesion equates to loss of the coating's protective function. The reader will benefit to understand the critical conditions

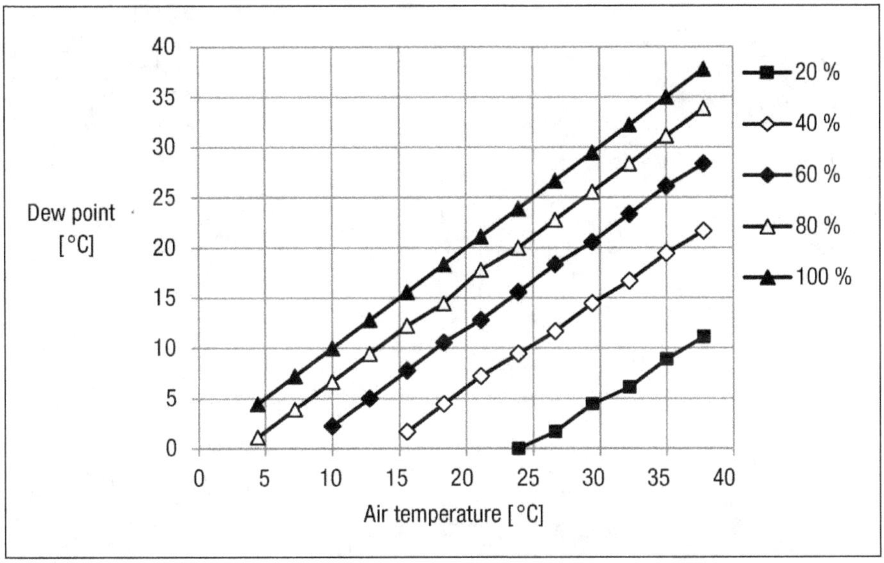

Figure 3.8: Dew point (T_d) as a function of air temperature at constant atmospheric pressure

that support a good adhesive bond between concrete and primer. For floor installation to both newly poured concrete and old concrete, mechanical abrasion is recommended [149, 150, 143].

Once a concrete foundation has been prepared for flooring installation, local temperature and humidity conditions and their changes before, during and after the installation are paramount for a successful job [151, 152]. The conditions and changes are best characterised by a phenomenon called dew point, which is illustrated in Figure 3.8. The dew point may be defined as the temperature – at constant atmospheric pressure and water-vapour content – to which the air must be cooled down to, to reach saturation [153]. In other words, at 100 % relative humidity, the actual temperature equals to the dew point. The condensed water is called dew when it forms on a solid surface.

Relative humidity in concrete usually remains quite high, typically about 75 %. This equates to a moisture content of about two per cent [154]. Furthermore, the concrete slab attracts moisture because it is almost always colder than the air temperature. If the concrete surface cools down, it may reach dew point and literally can be saturated with moisture. As a rule of thumb, conditions for flooring installation are favourable on a concrete slab with two per cent moisture, while surface temperature is five degrees Celsius above dew point and relative humidity is controlled at 60 to 70 % during application and curing.

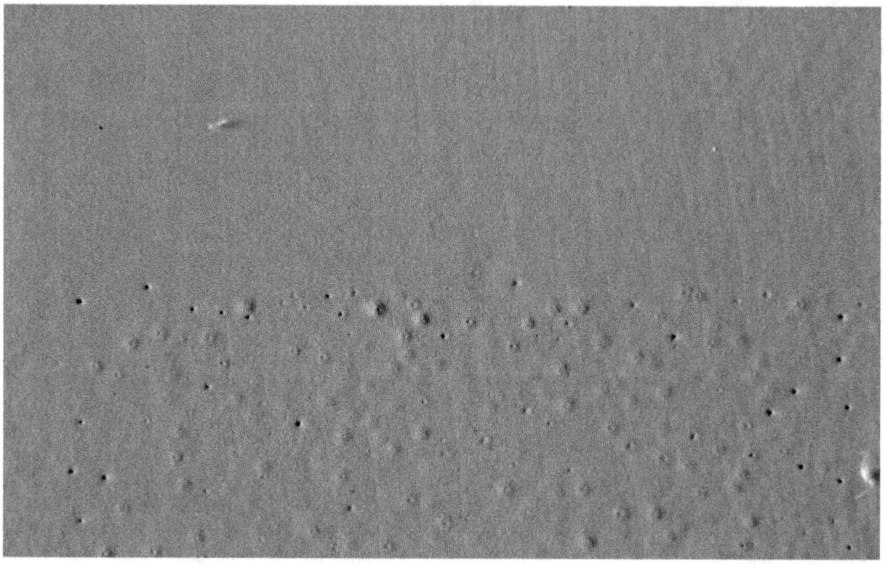

Figure 3.9: Comparison between floor application with a primer (top half) and without one (bottom half). Concrete's porosity causes issues with blistering and pin-holing which is resolved by use of a primer. Source: Air Products and Chemicals, Inc.

In spite of appropriate concrete preparation, delamination and blister formation of industrial floors are a re-occurring theme. It can be mitigated with the appropriate procedures and products in place. The discussion requires an understanding where moisture or water vapour originates from. Porous materials are a primary concern for blister formation and therefore best avoided or at least properly sealed [155, 156] (see Figure 3.9). Also, mixing errors on the job site favouring excess amines to epoxy resin may lead to coatings with higher water uptake [157] and problems later on.

For adequate quality concrete, consider the following means of moisture migration [158]. In wet conditions below ground level, concrete is potentially subject to a hydrostatic head, which drives water upward. Though recommended [159], in many existing situations it is unclear if a waterproofing membrane has been applied or is intact. Secondly, concrete can form so-called osmotic cells [160–162], which create pressure in excess of (low) adhesive bond strength and can cause osmotic blistering. Water vapour transmission (WVT) to the surface is yet another root cause where water vapour inside the concrete diffuses upward through the pores. For newly applied concrete, for example, the moisture feed of WVT originates from the cement hydration reaction. For floor installation to existing or old concrete, WVT may originate from temperature and humidity differences inside concrete and the room above. Various authors have described flooring solutions that mitigate the risk of osmotic blistering and WVT effects [162–165].

Carbamation resistance and appearance

Surface appearance is described by the ability of the floor coating to level and form a smooth surface. Gloss is typically a good method to characterise the surface appearance. Carbamation is a complementary parameter describing cured epoxy thermoset coating surfaces. It is derived from the amine side reaction with moisture and carbon dioxide to form salts (Equation 3.16). Moisture and carbon dioxide are typically present in the air. Concurrently, the porous nature of concrete may create similar conditions in the substrate when carbon dioxide is dissolved in water. Room temperature cure epoxy thermoset systems take time to develop a solid (cured) state, which means amine mobility is relatively high below T_g. As a result, side reactions take place at the film-air and film-substrate interfaces. The severity with which they occur is a function of the type of amine curing agent used [166].

Often, salt crystals formed are latent and only become visible as a white haze or blush upon contact with water. The phenomenon is known as water spotting and can have a significant impact on floor finishes and appearance. ISO 2812 provides a practical test method in which the coating or floor casting is cured under defined temperature and humidity conditions prior to applying a water saturated fabric for a specific duration, typically 24 hours. Early work showed the formation of salt crystals and their impact on cure speed and adhesion [167]. Further work investigated the nature of the salt and how the choice of amine curing agent can help formulators [168].

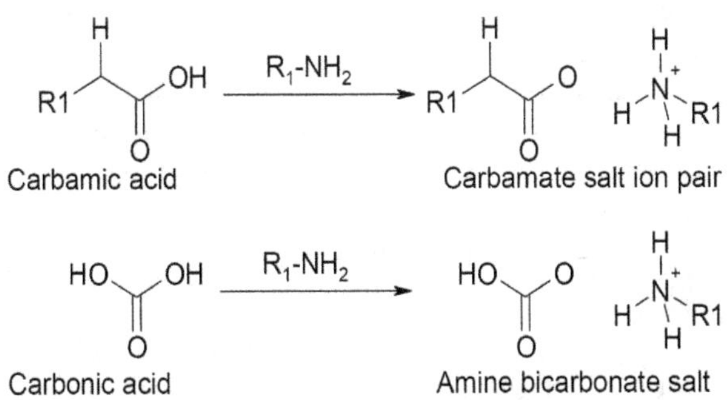

Equation 3.16: Chemical reactions leading to ammonium salts

Back in service time

Design of amine curing agent technology for high build floorings targets a balance between fast cure speed for quick back in service time and a sufficient potlife for (sub)ambient temperature application. The potlife is generally specified by the gelation time [169–171]; whereas rapid return to service is best defined by the time to reach a hardness of Shore D50 (ISO 868). The latter, also referred to as 'walk-on' time, is broadly accepted as a level of hardness for the flooring product

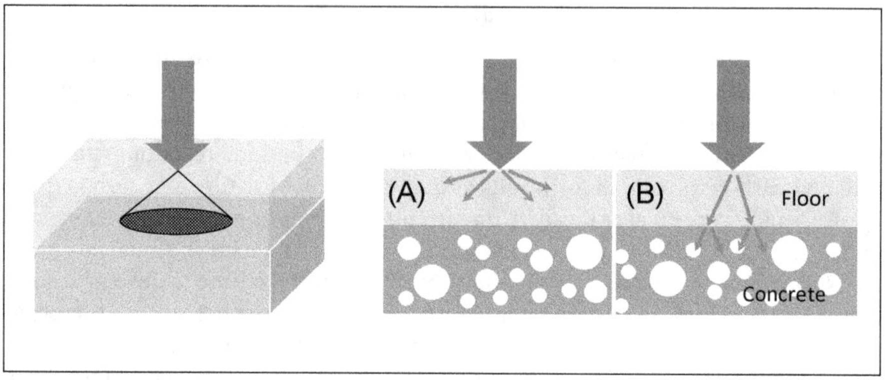

Figure 3.10: Compressive strength in industrial floors can be either low modulus (A) for load bearing property or high modulus (B) for monolithic behaviour.

Source: Air Products and Chemicals, Inc.

that allows return to service or application of a next layer. Additional data points for coatings such as drying times and pendulum hardness compile a tangible set of data that characterise the cure of the thermoset system as a function of time.

Mechanical resistance

Mechanical resistance forms the core of many epoxy thermoset based industrial floorings and provides elemental protection to concrete. In the extreme scenario's, a system provides either high modulus (monolithic) or a lower modulus for load bearing properties (Figure 3.10). The lower modulus absorbs force within the floor and reduces stress exposure to the concrete substrate. The opposite high modulus system, facilitates stress transmission throughout floor and substrate. Recent studies [172, 173] demonstrate how elimination of non-amine functional components in the epoxy thermoset can help facilitate high mechanical stability.

Complementary to compressive exposure is the ability of thermosets to absorb large amounts of energy without fracturing. Testing is typically carried out at a slow speed tensile mode to give elongation, or alternatively, at faster rates fracture mechanics, to measure resistance to crack propagation. The property is particularly relevant to provide continued protection to concrete slabs after cracks have occurred, such as in park decking and secondary containment. Epoxy thermosets typically have a brittle nature [174] and require modification with tougheners and flexible crosslinker components to meet the application requirements [175, 176].

Conductivity

As discussed in Chapter 2, fully cured epoxy thermosets generally provide high insulation properties. Though useful for many applications, certain operational conditions require a conductive system in order to avoid build-up of static electricity, or prevent electrical interference with sensitive equipment [177, 178]. Efficient electro static discharge (ESD) avoids generation of sparks and risks of explosion. Conductivity according DIN-EN 61340 is divided in three categories by their resistance to earth (RE): electrically conductive floors, dissipative floors and antistatic floor covers.

Wear resistance

During its service life, the floor system is exposed to various mechanical loads ranging from light foot traffic to heavy industrial use. As a result of these exposures, it is useful to characterise the resistance to wear. The floor's abrasion resistance (ISO 7784) is a practical indicator to characterise the wear resistance. In addition, scratch resistance (EN ISO 1518) provides a more subtle view of abrasion that impacts gloss and appearance of floorings in decorative application.

Chemical resistance

Exposure to chemicals and solvents is subject to the application and industry, which makes general guidelines difficult to provide [179–182]. In addition to the mode (immersion, patch resistance testing) and duration of exposure, resistance of the flooring products depends on factors such as binder cross linking, coating formulation specifics, as well as temperature and humidity conditions of importance to consider.

Temperature and UV resistance

The necessity of chemical modification and formulation of amine-cured epoxy thermosets (Chapter 3.3.2) introduces limits on the amenable maximum T_g. As a result, ambient cure epoxies typically display deflection temperatures between 40 to 50 °C. Though limited, the ambient operating temperatures in most industrial environments allow successful application of epoxy based floors.

Another limitation of epoxy thermosets is the UV resistance. DGEBA epoxy resin inherently introduces aromatic moieties in the epoxy binder, which typically contribute to higher yellowing of coatings. Industrial applications' focus on cost, adhesion strength, and chemical and mechanical stability are supporting the success. The trade-off on yellowing resistance is often accepted or compensated for by the application of an additional topcoat sealer.

Indoor air quality and emission compliance

At the European level progress had been made to improving safety and indoor air quality to inhabitants and users of buildings and domestic areas [183–185]. Legislation to manage potential emissions to inhabitants and environment throughout the life span of the floor is in place. Approaches in Europe are based on EN-ISO 16000 but vary per nation. German DIBt's AgBB protocol [186] attempts to limit emissions in cured floor products. A similar approach is followed in France and Belgium [187]. Swiss' Minergie [188] certification drives sustainable buildings based on reaching threshold levels in all key performance. For floorings this means emissions are controlled over the VOC of the final product. Leadership in energy and environmental design (LEED) is an initiative by the U.S. Green Building Council (USGBC) [189] and works through a rating system. Though only buildings can be LEED certified, low emission flooring products help to achieve points in the total assessment.

Emission reduction initiatives have introduced new challenges in the development of amine-cured epoxy thermosets, in particular, with the elimination of the plasticiser component(s) in incumbent products. Initial offerings were based on modified polyether diamines (Equation 3.15) and typically came with long walk-on times and reduced chemical resistance. Recent solvent-free initiatives and waterbased epoxies have closed the gap with incumbent technology. They often provide additional performance benefits to the coating manufacturer [172, 173, 164, 190].

3.3.5 High performance industrial flooring

A primer on concrete is the starting point for all polymer flooring installations. It serves to penetrate and seal the concrete pores and establishes good adhesive bonds with the concrete substrate and the polymer over-layment. The primer also provides a first protective layer to concrete against wear and tear and chemical exposure. Amine-cured epoxy thermosets have a track record for use in industrial primers and sealers on concrete. Often they are applied without additional fillers or additives at 200 to 500 g/m^2 using a squeegee. Typically the minimum application temperature is eight degrees Celsius while allowing a wide range of relative humidities.

Next to the primer, high performance industrial flooring installation typically requires one or two additional stages: the receiving coat and finish coat. These stages provide the flooring functionalisation as described for the key applications hereafter.

Light industrial and institutional flooring

The category of commercial and institutional floorings includes offices, hallways, canteens and areas for light industrial use. Performance requirements are aesthetics, cleanability and good wear resistance. High resistance to chemicals and mechanical load are often not required. In addition, there is an increasing demand for floorings to meet emission compliance for improved indoor air quality. Applying a thin coating up to 500 g/m^2 following the primer is a cost-effective way to

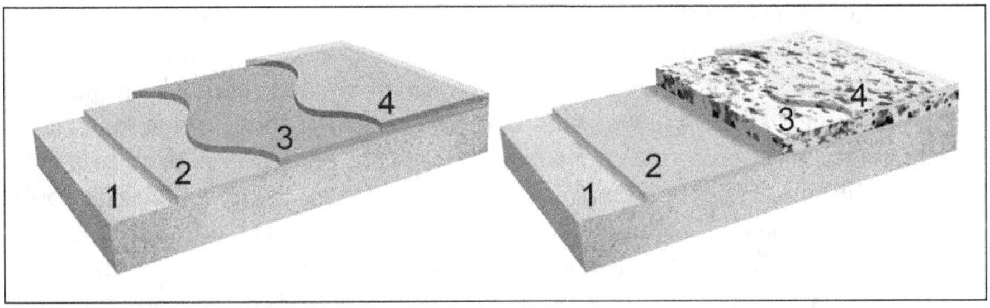

Figure 3.11: Design of high performance industrial flooring installation. The concrete substrate (grey) is coated with a primer (blue), a receiving coat and, finally, a transparent topcoat. Source: Air Products and Chemicals, Inc.

meet the objectives. Alternatively, a thin self-levelling floor as receiving coat adds value for decorative purposes when sprinkled with coloured chips. A transparent topcoat is optional and applied when satin-gloss finishes are preferred, alone, or in combination with improved UV resistance. Waterborne epoxy systems are particular popular in this area of application.

General industrial flooring

Flooring for larger scale industrial sites, factories, exhibition halls and laboratories are examples of general industrial use. The environment is typically a demanding one, where performance dominates over aesthetics. Attributes such as load bearing property, resistance to wear and tear and chemical exposures are paramount for floor longevity. Refurbishment of existing floors means downtime of business operations, hence, the pressure to deliver quick back in service times. Following an epoxy primer, a 2 to 5 mm epoxy self-leveller floor is most frequently used.

Figure 3.12: Examples of industrial flooring: clean room processing (left), car parking (middle) and laboratory (right) Source: Air Products and Chemicals, Inc.

Typically, a self-leveller floor consists of three components: Part A, a diluted epoxy resin formulated with colourants, finer filler components and additives; Part B, the amine curing agent; and Part C, the coarser fillers and sand, which make up the aggregate package. After mixing components at the job site, the material is free-flowing and poured onto the substrate. A serrated trowel is used to spread the material and set application thickness by the height of the teeth. The applicator then uses a spiked roller to accelerate de-airation and avoid surface defects during curing. As a result, self-levelling floors are efficiently applied up to several thousands of squared meters per day. The cured floor is often ready for exposure to traffic 24 hours after application and at the lower temperature condition.

Storage and logistic warehouses

In addition to general industrial flooring, warehouses and storage areas often require higher resistance to mechanical exposure than concrete can provide. Concrete's strength is typically limited to about 25 MPa. Continued exposure above this limit leads to cracking and has the risk of further deterioration to occur. A thin receiving coat eliminates concrete dust and shields from chemical exposure but lacks mechanical protection. Therefore, load bearing floors are more appropriate. Their application can facilitate mechanical loads up to 2 to 4 times concrete's strength [173].

Park decks and secondary containment

Car park deck floors and underground parking areas follow the general industrial use with amendments. Firstly, the primer is evaluated for through cure under adverse conditions and for adhesion under critical conditions. The primer is broadcasted with sand to mitigate potential impact carbamation may have on adhesion with the receiving coat. Secondly, the receiving coat is a 2 to 5 mm epoxy self-leveller formulated with a membrane or co-binder to deliver higher elongation and flexibility. Thus via modification an epoxy floor can facilitate good crack bridging property under static conditions down to -10 °C. A topcoat is optional and may be applied to improve wear resistance and to provide functional use.

Secondary containment flooring can be considered a cross-over type. Typically, these floors have special requirements in terms of flexibility and chemical resistance depending on the area of industrial use. Containment floors are also required to protect ground water.

ESD flooring (pharmaceutical, electronics, printing industry – high speed rotation)

ESD floors are formulated with special fibres to give the floor its conductivity. In contrast to the previous applications, the primer is also formulated with fibres.

Figure 3.13: Example of ESD floor specimen – both primer and self-leveller are formulated with conductive fibres; copper strips are applied for efficient electrical discharge. Source: Air Products and Chemicals, Inc.

Special copper strips are applied into the primer, wet-in-wet, for improved conductivity. After overnight cure, a 2 to 5 mm self-levelling floor is applied, again using conductive fibres. Application of ESD floors cover a range of areas, including clean rooms, electronics processing areas and printing industry where high speed rotation presses can cause risks with static electricity. Also areas with combustible dust are at risk when static electricity sparks can initiate explosion. In particular for clean rooms and electronics industry, floorings are required to be free of volatile materials. These include slow evaporating substances, which vapours can damage sensitive equipment.

3.4 Powder coatings

Ulrich Christ

Thermoset powder coatings based on epoxy resins have an enormous economic importance, since they cover a broad area of both functional and decorative applications. Similar to the properties of epoxy resins such characteristics as very good adhesion, very good corrosion protection, very good mechanical properties and excellent chemical resistance are observed for epoxy powder coatings as well. Therefore, epoxy resins belong to the most important class of binders for powder coatings since the beginning [195].

Powder coatings mainly based on epoxies are suitable primarily for indoor applications, because of chalking and yellowing. However, in multilayer structures they are used as very well-adhering and anti-corrosive primers, for example, in the car supply sector and in sophisticated corrosion protection.

Powder coatings with good weathering and light stability can be formulated by crosslinking of epoxy functional triglycidylisocyanurate (TGIC) or aromatic glycidyl esters with high molecular weight carboxy-terminated (acidic) polyesters. In this coating technology the epoxy component acts as a curing agent.

Polyaddition reactions and polycondensation reactions upon elimination of water or alcohols play the most important role as crosslinking reactions for thermoset powder coatings. In the case of the still relatively insignificant UV-curing powder

Table 3.23: Powder coatings based on epoxy binders or epoxy functions

Powder coating type	Epoxy function	Hardener type	Applications
Epoxy powder coating	EP resins based on BPA	Dicyandiamide (DICY), – substituted or – accelerated polyphenols, phenoxy resins, resols, novolaks; anhydrides	Heavy duty corrosion protection, pipe coating functional powder coatings, radiators, pipe coating, electrical and electronic applications, food and beverage containers
Hybrid or EP/PUR powder coating	EP resins based on BPA	Carboxy functional polyesters	Corrosion protection, decorative and universal use for industrial applications; MDF
Polyester powder coating	Glycidylesters, TGIC	Carboxy functional polyesters	High durable coatings for exterior applications, e.g. architectural coatings
Acrylic powder coating	Glycidylmeth-acrylate (GMA)	Dicarboxylic acids, e.g. DDDA	Automotive clear coats, high durable industrial coatings

coatings, curing takes place by radical-induced polymerization. Similar to liquid coating materials, in thermoset powder coatings the hardener components have a main significance in the formulation in addition to the binders. They determine key properties in the entire process chain such as extrudability, extrusion temperature, homogeneity of the dispersion, storage stability, curing temperature, flow properties, and the properties of the finished coating. Table 3.23 shows an overview of powder coating systems based on epoxy binders respectively epoxy functions.

Solid BPA-based epoxy resins with molecular weights of 1,500 to ~ 3,000 g/mole are used in powder coatings. In general, the powder coating or powder coating properties can be influenced as a function of the molecular weight as follows [196]:

Increasing molecular weight leads to:

- increased storage stability,
- declining in levelling properties,
- improved edge coverage,
- and increasing flexibility and impact resistance.

Increasing functionality of the reactants, and thus increasing degree of crosslinking leads to:

- decreasing flexibility and impact resistance,
- increasing chemical resistance,
- and increasing surface hardness.

The chemical resistance of epoxy powder coatings can be further improved by the modification of BPA-epoxy resins with polyfunctional glycidated phenol or cresol novolaks. This modification also results in the improvement of the electrical, chemical and mechanical properties at elevated temperatures, though partly at the expense of the flexibility and sometimes the adhesion of the coating [196].

Powder coatings based on BPA are crosslinked primarily with the following curing reactants [196–198]:

- dicyandiamide (DICY), accelerated or substituted,
- polyanhydrides,
- polyphenols,
- carboxy-functional polyesters.

In powder coatings with epoxide as the curing agent, glycidyl esters or triglycidyl isocyanurates with carboxy-functionalized polyesters are used. In the case of acrylic powder coatings, epoxy acrylate resins are cured with dicarboxylic acids. The following chapter gives an overview on epoxy powder coatings.

3.4.1 Epoxy powder coatings

The epoxy resins played the most important role in powder coating technology from the beginning. While their use was essentially limited to functional applications in the early decades, applications in the decorative sector opened up gradually. Nowadays epoxy and hybrid powder coatings provide more than 50 % of the powder coating market. For the manufacture of epoxy powder coatings, solid EP-resins are used, predominantly based on BPA, whose average molecular weights vary from about 1,500 to 3,000 g/mole, with sufficiently high T_g's in the range between 50 and 65 °C and melting in the temperature range between 80 and 100 °C. The curing process is carried out predominantly by polyaddition without any cleavage or condensation products, and thus, powder coatings with relatively high layer thicknesses without film defects can be produced [197].

The main types of curing agents are dicyandiamide (DICY), substituted or modified DICY, polyphenols, anhydrides or esters of polycarboxylic acids. The mechanisms of the respective crosslinking reactions of the powder coating systems are discussed in Chapter 2.

3.4.1.1 Curing with dicyandiamide (DICY)

The curing of epoxy resins with dicyandiamide (DICY) belongs to the type of epoxy-amine crosslinking. DICY is the most widely used curing agent for EP powder coatings.

Due to the high melting point (mp) (mp of unmodified DICY: 210 °C) and the low solubility of DICY in epoxy resins, epoxy powder paints have a very good storage stability. Because no melting occurs during the extrusion process due to the very high melting point, DICY must already be present in the pre-grinding process in the form of very fine particles in order to achieve a possibly homogeneous mixing with the epoxy resin.

For this, micronized, finely ground DICY powders, i.e. with particles down to 5 µm, are available from the raw material suppliers for powder coatings. To compensate for a possible proportion of inhomogeneous distribution in the extrudate and for the poor solubility in the epoxy resin, a slightly higher than the needed stoichiometric amount of DICY is used in the powder coating formulation. Since the reactivity of DICY cured epoxy powder coating materials is very low compared to amines in liquid epoxy coatings, and the dissolving in the epoxy resin, followed by the curing reaction that does not start until 185 °C, the baking conditions for the sheer DICY-curing are at 30 min/185 °C, 20 min/200 °C [199] or 15 to 30 min/220 to 185 °C [197].

The curing temperature may be significantly lowered primarily by two classes of accelerators: by substituted urea derivatives (urones) and by imidazole derivatives, which connect the high reactivity with good latency and good mechanical properties, including high T_g's of the films [200–203]. Uron-based accelerators are also distinguished by low yellowing and low toxicity. With these types of accelerators, the DICY curing can be adjusted to gradual reactivity, wherein curing conditions of e.g. 15 min/140 °C are possible without affecting the storage stability of the powder coating [200–203].

With chemically modified or substituted DICY systems, e.g. with Di-o-tolylguanidine (OTB) [200] and others, the dissolving temperature in the epoxy resin can be lowered to 60 °C. The mixing of the epoxy resin with the curing agent is thereby more homogeneous, and therefore more reactive powder coatings with baking conditions of 140 °C/15 min or 145 °C/8 min can be produced [200, 204], which lead to high-gloss and very flexible powder coatings, as shown by the example of the powder coating according to the guideline formulation given in Table 3.24 [200].

Owing to the availability of matt curing agents, the gloss range of epoxy powder coatings can be adjusted from high-gloss to matt with gloss levels of 16 (60° angle) [200, 205]. With chemically modified DICY and in particular due to the improved solubility of DICY derivatives, homogeneous coatings with a good coating levelling can be produced which no longer have the disadvantages of the films achieved with pure DICY, as well as the relatively high sensitivity to water [199].

Epoxy powder coatings with excellent adhesion, very high resistance against many chemicals including acids, alkalis and solvents, very good mechanical properties, such as abrasion, scratch and impact resistance can be made with substituted DICY using optimal formulations. Furthermore, the curing conditions can be flexibly adapted to technical coating and application process aspects.

Table 3.24: Guide formulation for a white glossy epoxy powder coating [200]

Pos.	Raw material/generic/trade name	Parts by weight
EP binder		
01	Epoxy resin based on BPA	60.0
02	Pigment	10.2
03	Filler	22.0
04	Levelling agent	4.0
05	Degasing agent	0.3
EP hardener		
06	Amine hardener	3.5
Sum		**100**
01	Epoxy resins, BPA type 3.5; EEW: 733 g/mol e.g. "Araldit" GT 7004 (Huntsman)	
02	Titanium dioxide rutile, e.g. Kronos 2160	
03	$CaCO_3$, e.g. "Omyacarb" SV1 (Omya)	
04	Levelling agent EP resin; EEW: 805 g/mol e.g. "Araldite" 2874 (Huntsman)	
05	e.g. Benzoin	
06	Substituted DICY e.g. o-Tolylbisguanamide, HEW: 18–21 ("Dyhard" OTB)	

Characteristic properties of powder coating

Curing conditions	140 °C/15min
Dry film thickness	65–70 µm
Gloss units (60° angle)	105
Erichsen cupping test [mm]	7.2
Ball Impact test [lbs]:	20

DICY-cured epoxy powder coatings are also suitable for the coating of outgassing substrates like die cast metals.

They show a good levelling and a good over baking stability in directly heated gas ovens. Another important application is the sophisticated corrosion protection. Coating systems with an epoxy powder primer on blasted steel and a polyester powder coating as a topcoat conform to the requirements of ISO 12944-6 in the corrosivity category C5-I high [206]. For the application of two-layer structures for the sophisticated corrosion protection, the epoxy base coat can be partly gelled at e.g. 160 °C/5 min, then top coated with a polyester powder coating and finally be cured together.

Adverse characteristics, as generally observed in coatings with aromatic epoxy resins, are the poor light resistance and the yellowing tendency, so that these coatings are preferably applied only for indoor use or as priming coats in multi-layer systems. Main applications of epoxy powder coatings are the corrosion protection in steel constructions and the coating of pipelines.

3.4.1.2 Curing with phenolic resins

An important group of curing agents for epoxy powder coatings are polyphenols from resoles and novolaks, which are produced for example from BPA or adducts of BPA in epoxy resins. Their chemical structure may be similar to the structure of the epoxy resins, or carry a terminal phenol or phenoxy group instead of a terminal epoxy function [197, 207, 208].

With the combination of phenolic resins based on resoles and novolaks or phenoxy resins with epoxy resins, a plasticizing of the phenolic resins takes place, which also has a favorable effect on the property profile of the coatings [195]. Basic catalysts, such as tertiary amines, imidazoles or quaternary ammonium salts, are used for crosslinking. Mostly they are already mixed with the curing agent, which promotes the achievement of a homogeneous mixing in the powder coating. Typical curing conditions with catalysts are 10 to 20 min/130 to 160 °C [197].

Epoxy powder coatings cured with phenolic resins are very highly crosslinked and therefore show very good resistance to chemicals, to solvents, and to hot mineral oils. They exhibit very good wet adhesion to metals even in the presence of boiling water and are extremely stable to hydrolysis. Their high gloss in conjunction with good levelling and the high storage stability (latency) of the powder coatings are also outstanding.

Due to the low colour stability and yellowing, the application is limited to functional areas where high thermal, mechanical, chemical, and electrical resistance is required in addition to excellent adhesion. Therefore, epoxy powder coatings cured with phenolic resins are ideal for applications in corrosion protection, such as the pipe coating, coating of hot-water radiators, coating of tanks, and electrical switch boards and control cabinets.

The good scratch and impact resistance also allow the coating of wire products, fire extinguishers, tools, electronic components, electrical cabinets, security boxes, and car fittings. Even higher crosslinked powder coatings are achievable from e.g. curing agents based on cresol novolak with epoxy cresol novolaks, enabling coatings with very high glass transition temperature, T_g, very high thermal and excellent chemical resistance [209]. These are particularly suitable for the production of high-temperature and corrosion-resistant coatings, e.g. for pipe coatings or insulating coatings for electrical and electronic applications, where high resistance to high humidity and to cyclic temperature/humidity loads is required [210].

3.4.1.3 Curing with anhydrides

As anhydride curing agents for epoxy powder coatings only polyanhydrides based on trimellitic or pyromellitic acids are used. The curing reaction of these types is usually accelerated by tertiary amines. Coatings based on this curing chemistry are highly crosslinked and therefore exhibit high solvent and chemical resistance, especially to acids. Such coatings are suitable for food and beverage containers [211] due to their neutral smell and taste.

As general properties, epoxy powder coatings adhere very well to metals, reveal a very good corrosion protection and a very good chemical and mechanical resistance due to high crosslinking densities. Their application range therefore includes many functional applications.

3.4.2 Epoxy polyester powder coatings or hybrid powder coatings

The term hybrid powder coatings is based on the crosslinking of epoxy resins with acidic polyester resins. The combination of epoxy resins with carboxy-functional polyester resins is in general one of the most important applications of thermosetting powder coatings. Hybrid powder coatings can be formulated on the basis of solid BPA-containing epoxy resins and carboxy-functional saturated polyester resins. The used polyester and epoxy resin components must have a glass transition temperature of 50 to 60 °C, so that a smooth milling of the extrudate is possible and the powder coating is storable without sintering. Linear or slightly crosslinked polyesters are mostly used, predominantly from the esterification of iso- and terephthalic acid with relatively short-chained polyols, e.g. neopentyl glycol, with relatively low molecular weights of 1,200 to 3,600 g/mole and acid numbers between 30 and 90 mg KOH/g [212–214]. The mass-related proportions between the polyester and the epoxy resin are between 70(80) : 30(20) and 50 : 50 [197, 212–215].

The baking conditions are from 130 to 200 °C/20 to 10 min. The crosslinking reaction takes place via the polyaddition of the carboxy group to build-up β-hydroxy esters upon ring-opening. Since the carboxy groups also catalyze the self-crosslinking of the epoxy resin, any excess of the epoxy resin can be used. The crosslinking reaction is catalyzed by conventional esterification catalysts, such as phosphines, phosphonium salts, tertiary amines, or metal salts [197].

Hybrid powder coatings are also suitable for special applications as so-called low-temperature or low-bake powder coatings. Low temperature curing is defined as curing at temperatures of ≤ 160 °C. One typical application is the coating of wooden materials such as MDF (medium density fibreboards) for the furniture industry. Here also hybrid powders with ultra-low temperature curing or ultra-low bakes (ULB) are used. The latter are defined by baking at about 130 °C/10 min [216]. In the case of 50 : 50 hybrids highly accelerated systems with a low glass transition temperature, T_g, and softening temperature of the polyester for acceptable levelling of the polyester and epoxy resins e.g. of the type 2.5 are needed.

While the powder coating technology of MDF is now safe to use, the possible outgassing of volatile components and negative influences on the film formation complicate the coating of veneers, due to the high proportion of heterogeneous natural wood.

Reports on a new technology of powder clear-coating of veneered chipboards can be found in the literature [217]. With a highly reactive, transparent powder coating based on epoxy-polyester (hybrid powder), which cures with accelerators at 150 °C/3 min or 130 °C/5 min, it is possible to coat veneered chipboards after grinding treatment and conditioning under controlled climatic conditions using corona discharge, and to cure them at 130 °C in a hot press under 64 N/cm^2. This new technology allows a surface finish which is applied in a single-layer process to a high-gloss clear powder coating (gloss level of 65 to 75 (20° angle)) with 100 to 140 μm thickness and having optimum film properties.

Another in-mold process, similar to the aforementioned method, allows to obtain high-gloss powder coatings on MDF materials with a well levelling, curing at low temperature pure epoxy powder coating [218]. Also for plastic components and metal parts that do not tolerate high temperatures for functional reasons, such as pre-assembled modules, there are numerous low temperature powder coatings with "reduced baking conditions" [219].

Table 3.25: Guide formulation of a white hybrid powder coating (60 : 40) [214]

Pos.	Raw material/generic/trade name	Parts by weight
01	Polyester resin	33.0
02	Epoxy resin based on BPA Type 3	24.0
03	Levelling agent	3.0
04	Pigment TiO_2 rutile	29.0
05	Filler $BaSO_4$	10.5
06	Degasing agent	0.5
Sum		**100.00**
01	Polyester resin, medium reactive, acid number: 60; T_g: 54 °C; e.g. "Crylcoat" 1620-0 (Allnex)	
02	BPA resin, type 3, e.g. "Epikote" EP 3003 (Momentive)	
03	Levelling agent, e.g. "Additol" P896 (Allnex)	
04	Titanium dioxide, rutile, e.g. Kronos 2160 (Kronos)	
05	Filler, e.g. Blanc fixe F (Sachtleben)	
06	e.g. "Benzoin" (BASF)	

Characteristic properties of powder coating

Ratio polyester/EP resin	60:40
Curing conditions	160 °C/20 min
Gloss units (20°/60° angle)	83–87/95–100

In general, hybrid powder coatings combine the properties of both binder components. Due to the polyester portion they show a better resistance to yellowing and chalking than pure epoxy systems.

Because of their optimal property profile, the ease of manufacture, the relatively low price, good all-round properties and universal applicability, the hybrid powder coatings are the most widely used powder coatings in Europe (proportions: 50 % in quantity). Although the main applications are substantially restricted to indoor use due to the lack of light resistance, these powder coatings are also suitable for outdoor areas such as for garden furniture, for priming coatings in outdoor applications with high requirements, e.g. used in vehicle construction for light-alloy wheels or for the corrosion protection of steel.

There are deficits compared to epoxy powder coating concerning the corrosion protection and resistance to chemicals. But these deficits are acceptable for a variety of applications in the decorative area, for applications in storage technology, office furniture and household appliances, for the construction sector, such as for ceiling panels, radiators, and even partially acceptable for outdoor use. A guide formulation for a white hybrid powder coating is given in Table 3.25 [214].

3.4.3 Polyester powder coatings

Polyester powder coatings are based on the crosslinking of carboxy-functional polyesters with light-stable di- or higher functional glycidyl esters or triglycidyl isocyanurates (TGIC). For this purpose, polyester resins having about the same or similar composition as the products described above for hybrid powder coatings, but with a higher molar mass between 3,000 and 7,000 g/mole, an acid number of about 23 to 30 mg KOH/g and glass transition temperatures of 62 °C to 70 °C, are used [212, 213, 220]. The glycidyl or glycidylisocyanurate function here as curing agents. Their substantially lower epoxide equivalent weights of about 100 g/eq(TGIC) or 150 g/eq (glycidyl ester PT910) in comparison to epoxy resins in hybrid powder coatings result in a mixing ratio of polyester to glycidyl ester of about 90 : 10 to 93 : 7 [212, 220, 221]. The crosslinking reaction takes place via the polyaddition of carboxy groups upon ring opening of the glycidyl oxirane under baking conditions of 160 °C/15 min to 200 °C/15 min [198, 220]. Influences of formulation modifications on the manufacturing parameters and coating properties are described in the literature [222, 223].

Polyester powder coatings which are based on this crosslinking chemistry are characterized by high light and weathering resistance, high chalking resistance, very good over baking stability, even in directly heated gas ovens, good chemical resistance, as well as by good mechanical characteristics such as hardness, flexibility and generally good processing properties [197, 198]. Due to the polyaddition curing reaction no condensates are produced, so that pore-free coatings without limitations of the layer thickness are available.

For high-quality outdoor applications, such as aluminium facades and windows, there are guidelines and quality regulations for the classification of polyester powder coatings according to their UV and weathering resistance, e.g. the GSB AL 631 directives [224] and the Qualicoat guidelines [225]. As an example, pursuant to GSB International, powder coating materials which perform with a gloss retention of ≥ 50 % in a test of three or five years under Florida weathering, are classified to be of Master or Premium quality. Polyester binders with a very high UV- and weathering resistance according to the above described criteria are often called superdurable polyester resins by the producers of raw materials. A guide formulation of a polyester powder coating, whose binder is suitable for highly resistant coatings, is given in Table 3.26. Polyester powder coatings with triglycidyl isocyanurate (TGIC) as the curing agent have been of greatest importance for highly weathering resistant powder coatings over a period of more than 30 years. Until a few years ago, TGIC-cured polyester powder coatings represented the global standard for highly weatherproof exterior coatings, for example, for the coating of aluminium facades and window profiles.

The assumption of mutagenic effects led to an early development of alternative curing agents. Thus, alternatives were already available in May 1998 when the

Table 3.26: Guide formulation of a white polyester powder coating [227]

Pos.	Raw material/generic/trade name	Parts by weight
01	Polyester resin, acid number: 23–27 mg KOH/g	64.80
02	Glycidylester; EW: 150	4.90
03	Levelling agent	1.00
04	Titanium dioxide rutile pigment	29.00
05	Degasing agent	0.30
Sum		**100.00**
01	Superdurable polyester resin, e.g. "Crylcoat" 4540-0 (Allnex)	
02	Crosslinking resin based on glycidylester e.g. "Araldite" PT 910 (Huntsman)	
03	Levelling agent e.g. "Modaflow" Powder 6000 (Allnex)	
04	e.g. Kronos 2160 (Kronos)	
05	e.g. Benzoin (BASF)	

Characteristic properties of powder coating	
Curing conditions	200 °C/10 min
Coating: dry film thickness on Al, chromated	60 µm
Gloss units (20°/60° angle), colour white, RAL9001	90/97
Gloss retention after weathering	
- Florida: 36 month 5° south: 60° angle	76 %
- QUV-B313: 600 h: 60° angle	83 %

TGIC was officially classified as a mutagenic substance. For about two decades, two substitutes for TGIC have been established on the European market:

- di- or triglycidyl esters of the terephthalic and isophthalic, tetrahydrophthalic and trimellitic acids (e.g. "Araldite PT 910" by Huntsman),
- β-hydroxyalkylamides (e.g. "Primid XL-552" by Ems-Chemie) or aromatic glycidyl esters [226] as hardeners for polyester powder coatings.

The aromatic glycidyl ester mixtures were developed to replace TGIC with a more favourable labelling while largely retaining the known beneficial properties after crosslinking with the acidic polyesters. "Araldite PT 910" has a slightly lower reactivity than TGIC due to a lower functionality. Powder coatings with > 1 % glycidyl of the solid content must be labelled as Xi (irritant).

3.4.4 Acrylic powder coatings

Acrylic powder coatings were developed in the 1970s, but were unable to penetrate the market initially. New developments of acrylic powder coatings in the last two decades belong to the most recent generation of products in the powder coating technology. They were designed to achieve clear-coat characteristics comparable to those of liquid coatings for automobile and automotive accessory applications. In particular, the levelling, the brilliance, the weathering and chemical resistance (especially against saponification) of the powder clear-coat systems should be improved and the curing should be possible at a relatively low temperature of 140 °C, suitable for the baking conditions in automotive coating technology.

Acrylic powder coatings can in principle be crosslinked via epoxide, hydroxyl and carboxyl functional groups. However, the most important method for the preparation of acrylic resins for powder coatings is currently the epoxy acrylate resin method. Thereby, epoxy acrylate copolymers of acrylic resins are produced in solution with specific proportions of co-monomer glycidyl methacrylate (GMA). Depending on the desired properties, such as glass transition temperature, T_g, melt viscosity, levelling, gelling time and crosslinking density, the monomers are selected and the acrylic resin with the desired molecular weight is produced.

The acrylic resins have epoxide equivalent weights between < 300 to > 1000 and molecular weights between < 3,000 to > 20,000 D. These highly functional GMA acrylic resins with glass transition temperatures of 42 °C to 58 °C are achievable with different reactivities and optimized for good levelling, or for example, for good pigmentability. For instance, a highly reactive GMA acrylic resin with good levelling can be used for automotive clear-coats and a medium-reactive GMA resin can be used for clear-coats on aluminium alloy wheels [228].

Acrylic powder coatings are produced using polycarboxylic acids and their anhydrides as crosslinking components. An important crosslinker is dodecane

Table 3.27: Guide formulation for a high gloss acrylic powder coating [230]

Pos.	Raw material/generic/trade name	Parts by weight
01	GMA-acrylic resin; EEW: 505	77.70
02	Dicarboxylic acid crosslinking resin, EW: 115	16.82
03	Levelling agent	2.17
04	UV absorber additive	1.89
05	HALS (hindered amine light absorber)	0.95
06	Degasing agent	0.47
Sum		**100.00**
01	GMA-acrylic resin, EEW: 505 g/Eq e.g. "Almatex" AP4411(Anderson Development Company)	
02	Crosslinking resin; EW 115 e.g. Dodecane dicarboxylic acid (DDDA)	
03	Levelling agent e.g. "Modaflow" Powder III (Allnex)	
04	UV absorber e.g. "Tinuvin" 405 (BASF)	
05	HALS, e.g. "Tinuvin" 144 (BASF)	
06	Degasing agent e.g. Benzoin (BASF)	

Characteristic properties of powder coating	
Curing conditions	163 °C/30 min
Dry film thickness	65–75 µm
Gloss units (20°/60° angle)	88/100
Appearance: wave scan: LW/SW/R-value	5.7/13.9/8.0
Impact resistance [inch-lbs]	35
Scratch resistance: Crockmeter; gloss units (20°angle)	26
Mandrel bending	ok
Chemical resistance to MEK: 100 double strokes	ok
Weathering resistance	
- Artificial weathering: QUV B-313: 2000 h	
- Gloss retention	80 %

diacid (C12). The crosslinking reaction is carried out by a polyaddition of the carboxy group of the dicarboxylic acid upon ring opening of the oxirane ring of the glycidyl epoxy acrylate resin, whereby a β-hydroxy ester is formed. The resulting hydroxyl group can further crosslink, for example, with an anhydride or with a capped polyisocyanate. Since these crosslinking reactions take place at temperatures between 140 °C and 160 °C, this technology is applicable for automotive clear coats.

The powder clear-coatings, which are made from epoxy acrylate resins and crosslinked with aliphatic carboxylic acids, satisfy typical requirements for automotive clear-coats such as very high weathering resistance, very good chemical resistance (especially against saponification), high gloss and brilliance (very good appearance) as well as hardness and elasticity [197, 228].

As compared to the current state of the art, acrylic powder coating materials have serious disadvantages in the processing at powder-coating facilities, since they are generally incompatible with polyester and hybrid powder coatings. Thus, to avoid surface defects caused by contamination, the application of acrylic powder coatings has to be carried out spatially separated from other powder coating systems. For these reasons, as well as due to the high cost of materials from the elaborate manufacturing process of the polycarboxylic acids or GMA acrylates, so far no breakthrough could be achieved – despite initially successful application as a powder clear-coating in the automotive industry. Successful applications of acrylic powder coatings are the clear-coating on light-alloy wheels [229], the top coating (pigmented) for an agricultural machinery manufacturer and until recently the clear coating used by an automobile manufacturer.

In recent years, the GMA-acrylic resins have been further developed with success in terms of improving the compatibility with polyester powder coatings and hence, more compatible acrylic powder coatings are currently in the market. Table 3.27 shows a basic formulation for a compatible acrylic powder coating [230]. Furthermore, there are also GMA-acrylic resins available for matting of acrylic powder coatings. Nowadays, acrylic powder coatings with a curing temperature of 145 °C can be made using appropriate GMA-acrylic resins, curing agents and catalysts.

3.5 Can and coil coatings

Michael Dornbusch

3.5.1 Can coatings

Glazes, waxes and bitumen have been used for several thousand years to seal containers.

The packaging industry has now developed into a key outlet for coatings and may be divided into cartons and tubes, which are primarily coated with printing inks, and cans. The latter can be sub-divided into cans for beverages and foods on one hand and barrels and non-food cans on the other.

Food cans are made from tin-coated steel (tinplate) or aluminium, and are coated on both sides, whereas barrels are produced from sheet steel [2, 11 and 12].

In the USA, beverage cans are made exclusively from aluminium, while both tinplate and aluminium are used in Europe. Although aluminium beverage cans are soft, they are held in shape by the relatively high pressure of the carbonic acid in carbonated drinks and beer. Food cans are made from tinplate because it would take a relatively thick and therefore expensive amount of aluminium to generate the necessary rigidity [11, 12].

There are two types of cans: 3-part and 2-part.

The 2-part cans are produced by means of a deep-drawing process in which the base and the body are produced in a single piece. This part is then combined with the prepared cap by deformation and joining processes.

In 3-part cans, one part is the body and the other two parts are the base and the cap [11, 12]. Smaller cans, especially beverage cans, are produced by a 2-part process while bigger food cans need to be made by the 3-part process.

Coating of 2-part cans

Deep-drawn cans are coated on the outside by means of a roller as they rotate and are then cured at 200 °C for several seconds, but only long enough to effect pre-curing. The interior coating is applied by spraying the rotating can, followed by curing the exterior and interior coating for about two minutes at 200 °C.

The high degree of automation possible and the fact that the cans rotate make it possible to coat 3,000 cans per minute. For an absolutely non-porous coating, each can needs 100 to 300 mg coating. Depending on the aggressiveness of the food, one- or two-layer systems may be applied [11, 12].

Coating of 3-part cans

Large food cans, in particular, can only be made as welded 3-part cans. Normally, coil coated (see Section 3.5.2) tinplate sheets are used, one sheet of which is used to make the cylinder by shaping and joining methods. Joining is effected either by welding or by adhesives. Soldering can only be used for non-food packaging due to the toxicity of the lead solder.

The hot surface of the weld is coated by spraying a rapid-curing coating to protect the metal surface. Solventborne coatings are generally used, although water-borne systems are available on the market. For very aggressive media, powder coatings find application and are becoming more and more popular [11, 12].

Technology behind the coatings

The composition of the interior and exterior coatings depends on the food to be packaged.

Not only must they be pasteurisable above 100 °C and sterilisable at 70 to 100 °C, but the coatings must also withstand cooking processes at 120 °C/60 minutes [12, 22]. Both coatings must offer permanent adhesion and present an intact surface after the wet and dry cycles. This treatment is best withstood by coatings based on organosols or epoxy resins cured with phenol-formaldehyde resins, because these systems have the best dry and wet adhesion before and after sterilisation. Electrochemical impedance spectroscopy (EIS) provides reliable verification of this adhesion [23-25].

Fruit and vegetables are normally packaged with coatings based on BPA-based phenoxy resins and crosslinked with a resol-phenol resin (see Section 2.4.1) in the presence of phosphoric acid as catalyst.

This coating is increasingly making way for transparent or white coating systems crosslinked with melamine or urea resins [2, 5]. An example of a formulation for an interior can coating is given in Table 3.28.

Colour changes during curing and the low weatherability are irrelevant in this application. The necessary flexibility is achieved by using high-molecular epoxy or phenoxy resins in combination with several complementary resins, such as phenol, melamine urea resins or isocyanates as crosslinking agents (see Section 2.4).

Modified phenol resins have now been developed that permit the use of low-molecular epoxy resins as a way to lower the solvent content (VOC) of the coating.

For cans of vegetables that release hydrogen sulphide during cooking, zinc oxide pigment is incorporated into the coating to react with the hydrogen sulphide and form white zinc sulphide and so avoid the production of unsightly black tin sulphide [2, 5, 12].

Using a mixture of epoxy resins and epoxy resins modified with phosphate esters (see Equation 2.20) can greatly enhance both the formability and the stability to lactic acid in the sterilisation test. It is worth noting here that only the mixture bestows this benefit – the use of resins modified with phosphate ester on their own does not [18].

Epoxy-phenol resins form the basis for particularly solvent-resistant coatings with a dry film thickness of 20 to 40 µm for coating drum interiors, which are increasingly being sprayed with waterborne coatings [11].

Table 3.28: Formulation for an interior can coating according to [48]

Mass.-%	Raw material	Characterisation
37.58	Epon 1001F	BPA based epoxide resin, EEW: 535
6.56	Acryloid AT-81	Acrylic resin with OH groups
29.41	Santolink KP 560	Butylated phenol-formaldehyde resin with 0.25 butoxy groups per phenol
0.32	WaCure 155	55 % solution of dinonylnaphthaline disulphonic acid in isobutanol
6.56	Butoxyethanol	
8.95	1-Butanol	
2.44	Phenol	
2.82	2-Propanol	
5.36	Water	
100	Total	

The waterborne coatings were specially developed from epoxy-phenol-formaldehyde emulsions. If conventional emulsifiers were to be used (see Section 2.4.3), storage stability in water would be significantly reduced. This is avoided by using amine salts based on poly(meth)acrylates, such as methacrylic acid/methyl methacrylate/ethyl acrylate/styrene polymers.

Epoxy resins modified with carboxylic acid may also be used; these are generated by esterifying the epoxy group with an acrylic acid, followed by neutralisation with an amine. The resins have properties similar to those of conventional epoxy phenol formaldehyde resins, unlike the case for formulations containing non-ionic emulsifiers [5, 12].

UV-curable coatings (see Section 3.1.2) are also used for coating beverage cans. The exterior coating resins are based on polyacrylates, which are cured by UV-initiated, free-radical polymerisation. The exterior coatings for the base and the cap are epoxy-based coatings which are partially cured by UV-initiated cationic polymerisation. The interior coating is a conventional epoxy/phenol resin, which is then thermally cured. The exterior base and cap coatings undergo thermal curing by residual Brönsted acids in the film and this further improves adhesion [12].

Toxicology of coatings based on bisphenol A (BPA)

Resins based on bisphenol (BPA) can possibly release BPA from the coating into the food or beverage. BPA is still present, albeit in low concentration, after resin production (see Equation 2.40 and 2.41). Daily intake of BPA-contaminated food ranges from 0.1 to 1.5 μg/kg [17].

The endocrinal effect in humans, e.g. BPA's oestrogenic activity (see Section 1.1) has been the subject of several studies in recent years [5] which focused on the extraction of BPA from packaging by food.

These showed that BPA migrates through the interior coating, several micrometres thick, of food and beverage cans during the sterilisation process. As different epoxy resins are used in different coatings, the quantity of BPA, which migrates into food varies over a wide range, but normally lies in the ng/g range.

More than 40 different compounds have now been detected that migrate into food; these constitute 50 % of the total migrated quantity [17].

It is for this reason that food and beverage cans need to be approved by food control authorities. In the USA, the Food and Drug Administration (FDA) has published tables, such as 21CFR175.300 (Resinous and Polymeric Coatings), that specify which compounds are acceptable in packaging [11, 12].

EU regulation 1282/2011 entitled "Commission regulation on plastic materials and articles intended to come into contact with food" is increasingly being applied to can-interior coatings [12], and lists migration limits for melamine, N-methylpyrrolidone, amines, alcohols and perfluorinated compounds and other common compounds and polymers.

The 2002 version (Directive 2002/72/EC) specifies a migration limit for BPA of 3 mg/kg (food) while Directive 2002/16/EC on "Materials containing certain epoxy derivatives and coming into contact with foodstuffs" imposes a migration limit for DGEBA and its hydrolysis products of 1 mg/kg.

DIN EN 13130-1:2004 and DIN EN 15137:2006 specify methods for studying the migration quantity of bisphenol A and novolak-glycidyl ester and their hydrolysis products. They describe the methods for ensuring compliance with Directives 2002/16/EC and 2004/13/EC.

At European level, there is the "European Union Risk Assessment Report, Bisphenol A, Part 2 Human Health", which dates from 2008 and was compiled by the UK in the context of REACH (Registration, Evaluation, Authorisation and Restriction of Chemicals). This report states that BPA does not constitute a hazard for human beings if the limit values mentioned in the directives above are observed.

As the results of the studies mentioned in this report are controversial, especially as regards the endocrinal effect, the 2008 results obtained as part of the REACH process are to be reviewed.

The controversy is reflected in Directive 2011/8/EU, which prohibits the use of BPA in baby bottles. This directive is based on studies from Denmark and France, in which a toxic effect was observed even at low concentrations. Even within this directive, the different aspects concerning the toxicity of BPA are highly controversial.

The European Food Safety Authority (EFSA) is currently preparing a new risk report on BPA contamination that will especially address food and the consequences for human beings [19].

In January 2015, the EFSA lowered the limit value for BPA from 50 µg/kg per day to 4 µg/kg/day [23], thereby approaching the values given in [17]. This remains a preliminary value for as long as the results from the long term rat study are outstanding.

3.5.2 Coil coatings

The first coil coating process began in 1935 with the production of Venetian blinds. It consisted in coating 5-cm-wide metal strips at a speed of 10 m/min. About 50 years ago, the coil coating industry increased the coating width to 200 mm. Modern conveyor systems can coat a 1.8 m wide coil at a speed of 275 m/min [11, 12, 27].

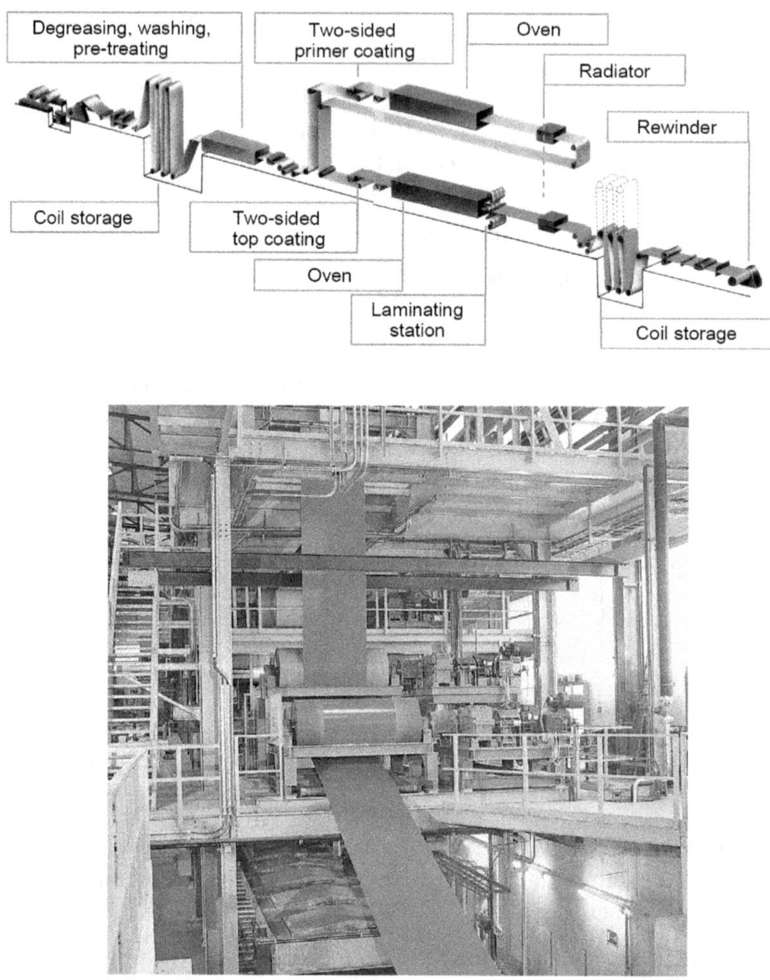

Figure 3.14: Schematic structure of a coil-coating line *Source: BASF Coatings*

The substrates for this process are steel, metal-coated steel and wrought alloys of aluminium. The metal-coated steel may be a hot-dip galvanized steel, galvanized steel, and alloys such as zinc-aluminium, aluminium-zinc, and zinc-magnesium.

Some ten years ago, zinc-magnesium coatings were developed that offer far superior corrosion-protection properties combined with a reduction in metal-coating thickness and unchanged corrosion-protection performance.

Coils between 0.15 and 3 mm thick and up to 1,800 mm wide are used for steel and up to 20,000 mm wide for aluminium, with conveyor speeds averaging between 80 and 140 m/min for steel and 150 m/min for aluminium [11, 26, 27].

Most coil coating lines for steel and metal-coated steel apply two organic coatings to the front and one to the back, but aluminium is generally coated on one side only.

The two-layer system consists of a primer, which has a dry film thickness of 5 to 8 μm, is applied by roller and is cured in an oven. The topcoat is also applied by roller, in a dry film thickness of 15 to 25 μm (up to 200 μm in the case of plastisols) and is cured in a second oven.

One-layer systems are also used and have a dry film thickness of 5 to 15 μm.

The coating on the back is applied at the same time as the topcoat and cured in the same oven as the topcoat [11, 12, 26, 27]. In view of the high conveyor speeds, flash-off must occur within 3 to 10 seconds after application. This imposes very high demands on film formation.

Curing occurs in continuous driers, which, due to the high conveyor speed, are extremely long and reach peak metal temperatures of 250 °C and offer curing times of 1 to 2 minutes [11].

Numerous classes of resins are used for topcoats. For façade engineering, plastisols such as polyvinyl chloride (PVC) and polyvinylene difluoride (PVDF) find application. A major group of resins for topcoatings is polyester resins that contain melamine-formaldehyde resins as crosslinking agents. Enhanced weatherability is afforded by silicone-modified polyesters.

Also employed are silicone-modified polyacrylates, polyacrylates, PUR with melamine-formaldehyde resins as crosslinking agents, and polyester crosslinked with blocked isocyanates [11, 12].

The resins for the primer normally consist of BPA-based epoxy resin polyester mixtures or epoxy ester crosslinked with melamine-formaldehyde resins. This combination produces good adhesion and outstanding corrosion protection.

PUR, latex primers and pure polyesters are used as well.

Epoxy-resin-based coatings normally contribute better corrosion protection than polyester-based primers, possibly due to the good adhesion of the epoxy resin to the substrate [11, 12, 16].

Table 3.29 shows a typical formulation for a conventional coil coating primer.

In the early days of coil coating technology, primers containing chromate were applied over a chromate-containing pre-treatment to provide outstanding corrosion

Table 3.29: Conventional coil coating primer, according to [46]

Mass.-%	Raw marterial	Characterisation
Grinding		
0.4	Silica	
0.5	Carbon black	
0.1	Titanium dioxide	
1	Talc	
9	Barium sulphate	
9	Zinc phosphate	
14	60 % Polyesterdiol in Solvesso 150*	OH number: 25 mg KOH/g, Mw: 5000
6	60 % Polyesterpolyol in 3:1 Solvesso 150 and butoxyethanol	OH number: 35 mg KOH/g, Mw: 4000
8	Dibasic ester	Mixture of dicarboxylic acid dimethyl esters (dimethyl adipate, dimethyl glutarate, dimethyl succinate)
1.7	Solvesso 100	Mixture of aromatic hydrocarbons in the boiling range 155–185 °C
Completion		
0.8	Common levelling agent	
1	Dibutyltin laurate	
22.5	60 % Polyesterpolyol in 3:1 Solvesso 150 and butoxyethanol	OH number: 35 mg KOH/g, Mw: 4000
8.4	75 % Butanonoxime-blocked isocyanate based on HDI	NCO content approx. 11,1 %
4.4	92 % Styrene/allyl alcohol-modified melamine resin	Hexamethoxymethylmelamine type
2.2	75 % Bisphenol A epoxide resin in xylene	EEW: 485
11	Solvesso 100	
100	Total	

* Mixture of aromatic hydrocarbons in the boiling range 182–202 °C

protection. For 20 years, chromate-free primers have been commercially available which, for 10 years, have been applied over chromate-free pre-treatments and offer corrosion-protection performance on a par with the chromate systems [13].

In contrast, water-borne primers containing epoxy resin and applied over a pre-treatment containing Cr(III) exhibited reduced corrosion protection compared with conventional chromate-primers applied over a chromate pre-treatment [15].

Table 3.30: Formulation for a water-borne coil coating welding primer, according to [47]

Mass [g]	Raw material	Characterisation
36.23	"Cymel" 303	Methylated melamine resin
202.03	Phenoxy resin modified with 31 % phosphoric acid, as per [47]	Acid number 18.1
292.42	Ferrophos HRS-3095	Iron phosphide
32.58	"Shieldex" AC3	Precipitated silica
0.91	"Surfynol" 104 PA	Defoamer based on acetylene diol
2.72	Levelling agent QR-708	

The turn of the millennium saw the launch of a number of research and development activities aimed at using coil-coating technology in the automotive industry. One approach favours an epoxy-based, zinc-dust primer applied in a dry film thickness of 2 to 4 µm over zinc-coated steel. The primer is formable and weldable and so can serve as a substrate in the automotive industry [14].

Table 3.30 shows the formulation for a water-borne, weldable primer, with iron phosphide serving as weldable pigment.

The back-side coating is a combination of an oxidatively-drying alkyd resin and melamine-formaldehyde resins, although the alkyd resin has been replaced by polyester in the meanwhile.

The back-side coating can be pigmented or a clearcoat, but, either way, contains an incompatible wax to avoid frictional damage to the coating during roll-up [12].

3.6 Literature

[1] J. Möckel, U. Fuhrmann, Epoxidharze, Die Bibliothek der Technik Vol. 51, Verlag Moderne Industrie AG & Co., Landsberg/Lech, 1990

[2] H. Kittel, Lehrbuch der Lacke und Beschichtungen, Vol. 2, 2nd Ed., S. Hirzel, Stuttgart, 1998, p. 267–318

[3] H. Kittel, Lehrbuch der Lacke und Beschichtungen, Vol. 3, 2nd Ed., S. Hirzel, Stuttgart, 2001, p. 38–47

[4] Encyclopedia of Polymer Science and Engineering, Vol. 6: Emulsion Polymerisation to Fibres, Manufacture, John Wiley & Sons, New York, 1986, 322–382

[5] Z. W. Wicks, F. N. Jones, S. P. Pappas, D. A. Wicks, Organic Coatings 3th Edition, Wiley Verlag, 2007, 271–294

[6] Z. W. Wicks, F. N. Jones, S. P. Pappas, D. A. Wicks, Organic Coatings 3th Edition, Wiley Verlag, 2007, 535–547

[7] J. V. Crivello, J. H. W. Lam, J. Polym. Sci., Polym. Lett. Ed. 17 (1979) 759

[8] E. Klemm, H.-J. Flammersheim, R. Martin, H.-H. Hörhold, Angew. Makromol. Chem. 135 (1985) 131–138 (No. 2173)

[9] EP000000259726A2
[10] A. Goldschmidt, H.-J. Streitberger, BASF Handbook Coating Technologies, Vincentz Network, Hannover, 2007, p. 105–110
[11] A. Goldschmidt, H.-J. Streitberger, BASF Handbook Coating Technologies, Vincentz Network, Hannover, 2007, p. 723–747
[12] Z. W. Wicks, F. N. Jones, S. P. Pappas, D. A. Wicks, Organic Coatings 3th Edition, Wiley Verlag, 2007, 574–618
[13] A. Heimes-Scheller, Asia Pacific Coat. J. 18, 6 (2005) 18
[14] M. Alsmann, JOT, 38, 10 (1998), 72–76
[15] M. Zapponi, Corr. Eng. Sci. Technol. 44, 2 (2009) 119–127
[16] F. Deflorian, L. Fedrizzi, P.L. Bonora, Electrochem. Soc. Proc. 97, 41 1998, 45–56
[17] L. K. Ackermann, G. O. Noonan, T. H. Begley, E. P. Mazzola, Rapid Commun. Mass Spectrom. 2011, 25, 1336–1342
[18] J. L. Massingill, R. C. Whiteside, J. Coat. Technol. 65, 824 (1993), 65–71
[19] www.efsa.europa.eu
[20] N. Yonet, N. Bicak, Y. Yagci, Macromolecules 2006, 39, 2736–2738
[21] N. Yonet, N. Bicak, M. Yurtsever, Y. Yagci, Polym. Int. 56 (2007) 525–531
[22] Dosen-ABC, Initiative Lebensmitteldose, München, www.initiative-lebensmitteldose.de
[23] M. Beiro, Prog. Org. Coat. 2003, 46, 97–106
[24] J. M. McIntyre, H. Q. Pham, Prog. Org. Coat. 27, 1996, 201–207
[25] A. C. A. de Vooys, B. Boelen, D. H. van der Weijde, Prog. Org. Coat. 73, 2012, 202–210
[26] Steel Information Centrum, Charakteristische Merkmale 093, Organisch bandbeschichtete Flacherzeugnisse aus Stahl, Ausgabe 2012, ISSN 0175-2006
[27] ThyssenKrupp Stahl AG, Bandbeschichtetes Feinblech, Juli 2005, Bestellnummer 2090
[28] DE000003727382A1
[29] DE102008061048A1
[30] US 7473324B2
[31] US000004859721A
[32] O. L. Abu-Shanab, W. E. Fristad, 2008 SAE International, Paper Number; 08-M66
[33] WO002012174424A2
[34] T. Laue, A. Plagens, Namen- und Schlagwort-Reaktionen in der Organischen Chemie, Teubner Verlag, Stuttgart, 1994, p. 110–113
[35] H.-J. Streitberger, K.-F. Dössel, Automotive Paints and Coatings 2nd Edition, Wiley-VCH, Weinheim, 2008, p. 3f, p. 99 and 241–257
[36] EP000000404033B1
[37] EP000000001431A1
[38] EP000000304854A2
[39] EP000000319834A2
[40] EP000000273170A1
[41] WO002009106337A1
[42] EP000000251772A2
[43] EP000000158128A1
[44] EP000000227942A2

[45] H. R. Christen, F. Vögtle, Organische Chemie Vol. I, Salle+Sauerländer, Frankfurt am Main, 1988, 655–657

[46] DE19612898C1

[47] DE60203696T2

[48] EP0458772A2

[49] Einblicke. Die deutsche Lackindustrie in Zahlen, Deutsches Lackinstitut GmbH (DLI), Frankfurt/Main; Ausgabe 2008

[50] Verband der deutschen Lack- und Druckfarbenindustrie e. V. (VdL), Kennzahlen der mittelständische Lack- und Druckfarbenindustrie in Deutschland 2013, http://www.lackindustrie.de/Themen/Marktinformationen/Seiten/Uebersichtsseite, aspx (2014-11-10)

[51] Internationale Qualitätsrichtlinien für Beschichtung von Bauteilen aus Aluminium GSB AL 63, Stahl und feuerverzinktem Stahl GSB ST 663, Ausgabe Mai 2013 GSB International e. V. Düsseldorf

[52] D. Stoye, W. Freitag, Lackharze. Chemie, Eigenschaften und Anwendungen. Carl Hanser Verlag München, Wien, 1996, p. 278

[53] D. Stoye, W. Freitag, Lackharze. Chemie, Eigenschaften und Anwendungen, Carl Hanser Verlag München, Wien, 1996, p. 264 (table)

[54] D. Stoye, Paints, Coatings and Solvents. VCH Verlagsgesellschaft Weinheim, 1993, p. 71

[55] Kunstharze Epikote-Ester Herstellung und Anwendung A-IV-1, Deutsche Shell Chemie Gesellschaft mbH, Frankfurt/M., Oktober 1974

[56] D. Stoye, W. Freitag, Lackharze. Chemie, Eigenschaften und Anwendungen. Carl Hanser Verlag München, Wien, 1996, p. 270–273

[57] D. Stoye, Paints, Coatings and Solvents, VCH Verlagsgesellschaft Weinheim, 1993, p. 72

[58] H. Kittel, Lehrbuch der Lacke und Beschichtungen, Vol. 2, 2nd Edition, S. Hirzel Verlag Stuttgart Leipzig 1998, pp. 291–295

[59] Caparol – Technical Information No. 091 – April 2013, CAPAROL Farben Lacke Bautenschutz GmbH, Ober-Ramstadt

[60] Kunstharze Epikote A-II-2, Lösungsmittelhaltige Epikote-Reaktionslacke Härtung mit Polyisocyanaten. Deutsche Shell Chemie Gesellschaft mbH, Frankfurt/M., September 1973

[61] D. Stoye, W. Freitag, Lackharze. Chemie, Eigenschaften und Anwendungen. Carl Hanser Verlag München, Wien, 1996, pp. 268–269

[62] D. Stoye, W. Freitag, Lackharze. Chemie, Eigenschaften und Anwendungen. Carl Hanser Verlag München, Wien, 1996, pp. 261–263

[63] Kunstharze Epikote A-I-1, Allgemeines über die Eigenschaften, Verarbeitung und Anwendung, Deutsche Shell Chemie Gesellschaft mbH, Frankfurt/M., January 1975, p. 18

[64] Kunstharze Epikote A-III-2, Epikote/Aminoharz-Systeme für Einbrennlacke, Deutsche Shell Chemie Gesellschaft mbH, Frankfurt/M., November 1972

[65] Allnex Startformulierung IB 418/2m Klarlackformulierung, Allnex, Graz, November 2014

[66] Kunstharze Epikote A-III-1, Epikote/Phenolharz-Systeme für Einbrennlacke, Deutsche Shell Chemie Gesellschaft mbH, Frankfurt/M., September 1972

[67] F. Bayer, A.W.H. Capell, O. Nicolai, J. Pflugfelder, Korrosionsschutz von Stahlbauten durch Beschichtungssysteme, Verband der Deutschen Lack- und Druckfarbenindustrie e.V., Bundesverband Korrosionsschutz e.V., 2010

[68] ISO 12944, Paints and varnishes – Corrosion protection of steel structures by protective paint systems Part 4: Types of surface and surface preparation, International Organization for Standardization, Geneve, Switzerland 1998

[69] ISO 12944-4:1998, Paints and varnishes — Corrosion protection of steel structures by protective paint systems Part 4: Types of surface and surface preparation. International Organization for Standardization, Geneve, Switzerland 1998

[70] Korrosionsschutz durch Duplex-Systeme (Feuerverzinken und Beschichten), Institut Feuerverzinken GmbH Düsseldorf, 2012; beginning page 15: Verbände-Richtlinie Korrosionsschutz von Stahlbauten Duplex-Systeme, gemeinsam herausgegeben vom BVK, VdL, DSTV und Industrieverband Feuerverzinken, Düsseldorf, 2000

[71] R M. Hernandez, J. Genesca, J. Uruchurtu, F. Galliano, D. Landolt, Effect of an inhibitve pigment zinc-aluminum-phosphate (ZAP) on the corrosion mechanism of steel in waterborne coatings, POC 56 (2006) 199–206

[72] G. Grundmeier, K. J. Roschmann, Wie funktionieren Zinkphosphat-pigmentierte Korrosionsschutzbeschichtungen, Farbe u. Lack 112 (2) (2006) 24–26

[73] J. Sinko, Challenges of chromate inhibitor pigments replacement in organic coatings, POC42(2001)267–282

[74] Schauer, L. Dulog, Effizienz von Korrosionsschutzpigmenten in Grundierungen, Farbe u. Lack 103 (1)(1979) 28–39

[75] A. Seth, W. J. van Ooij, P. Puomi, Z. Yin, A. Ashirgade, S. Bafna, C. Shivane, Novel, one-step, chromate-free coatings containing anticorrosion pigments for metals – an overview and mechanistic study, POC 58(2007) 136–145

[76] Heubach Anticorrosive Pigments, 02/2014, Heubach Langelsheim

[77] R. Naderi, M.M. Attar, Cathodic disbondment of epoxy coating with zinc aluminum polyphosphate as a modified zinc phosphate anticorrosion pigment, Progress in Organic Coatings, 69 (2010) p. 392–395

[78] R. Naderi, M. M. Attar, Effect of zink-free phosphate-based anticorrosion pigment on the cathodic disbondment of epoxy-polyamide coating, Progress in Organic Coatings, 77 (2014) p. 830–835

[79] M. J. Gimeno, S. Chamorro, R. March, E. Oro, P. Perez, J. Gracenea, J. Suay, Anticorrosive properties enhancement by means of phosphate pigments in an epoxy 2k coating. Assessment by NSS and ACET, Progress in Organic Coatings 77 (2014) 2024–2030

[80] S. Bender, M. Babutzka, L. Kirmaier, Korrosionsschutz schnell untersuchen, Farbe u. Lack 120 (5) (2014), p. 22–28

[81] Nubiola Nubirox Technical Guide 1.Ed., Barcelona, January 2009

[82] J. A. Vogelsang, Wasserhaushalt von Beschichtungen unter Temperaturwechselbedingungen. GfKORR-Jahrestagung 9 to 10 November 2010, Frankfurt/M., p. 68–76

[83] U. Christ, AiF-Forschungsvorhaben 17239N, Neue hochwertige Korrosionsschutzbeschichtungen für Stahl aus wasserverdünnbaren Polymerdispersionen, Schlussbericht 2014

[84] T. Schauer, H. Greisiger, S. Krieg, Eisenglimmer als Barriere. Farbe u. Lack 106 (2000) p. 73
[85] J. A. Vogelsang, U. Eschmann, G. Meyer, W. Strunz, Eisenglimmer in Korrosionsschutzbeschichtungen, neue Erkenntnisse zum Wirkmechanismus. Farbe u. Lack 104 (1998) p. 28
[86] TL/TP-KOR-Technical Terms of Delivery and Technical Test Regulations for Coating Materials for the Corrosion Protection of Steel Structures, 2002 Edition (TL/TP-KOR-Steel Structures) 2002 Edition (TL/TP-KOR-Steel Structures), Verkehrsblatt-Verlag Dortmund
[87] TL/TP-KOR Annex E, Blatt 87 EP/PUR, TL/TP-KOR-TL Sheets in: Technical Terms of Delivery and Technical Test Regulations for Coating Materials for the Corrosion Protection of Steel Structures, 2002 Edition (TL/TP-KOR-Steel Structures),Verkehrsblatt-Verlag Dortmund, pp. 59–64
[88] P. A. Sørensen, S. Kiil, K. Dam-Johansen, Anticorrosive coatings: a review, J. Coat. Technol. Res., 6 (2) 135–176, 2009
[89] M. Gerlitz, High Solid- und wasserverdünnbare EP-Beschichtungsstoffe für den Korrosionsschutz von Stahlbauten. U. Christ, Korrosionsschutz durch Beschichtungen, Technische Akademie Esslingen, 10th November 2009
[90] M.A. Epting, New raw materials for corrosion-resistent coatings, American Paint & Coatings Journal Jan. 17, 1994, p. 52–61
[91] H. Kittel, Lehrbuch der Lacke und Beschichtungen, Vol. 2, 2nd Ed., S. Hirzel Verlag Stuttgart Leipzig 1998, p. 300
[92] J. Pflugfelder, High Solid-Beschichtungsstoffe für den Stahlbau, in W. Ph. Öchsner, Korrosionsschutz durch Beschichtungen, Technische Akademie Esslingen, 13rd November 2007
[93] Die VOC-Richtlinie in Deutschland. Dokumente zu Lacken und Farben, Heft 8, Deutsches Lackinstitut GmbH, Frankfurt/M. (2002)
[94] Council Directive 1999/13/EC of 11 March 1999 on the limitation of emissions of volatile organic compounds due to the use of organic solvents in certain activities and installations
[95] Directive 2004/42/EC of the European Parliament and of the Council of 21 April 2004 on the limitation of emissions of volatile organic compounds due to the use of organic solvents in decorative paints and varnishes and vehicle refinishing products and amending Directive 1999/13/EC ("the Paints Directive") aims to prevent the negative environmental effects of emissions of volatile organic compounds (VOC) from decorative paints and vehicle refinishing products
[96] W. Marquardt, Flüssige Epoxidharze auf Basis von Bisphenol F, Farbe u. Lack 86, 1980 (5), p. 413–419
[97] B. Dreher, Lösungsmittelfreie Beschichtungen auf der Basis von Epoxidharzen, Chemische Rundschau 30 (1977), No. 48, p. 23–25
[98] Huntsman Advanced Materials Specialty Components Selector Guide 02.2014
[99] Huntsman Advanced Materials – Starting Formulation for a primer for heavy corrosion protection 2014
[100] Huntsman Advanced Materials Product Technical Information Araldite PZ 3961-1
[101] G. Hobisch, Wässrige 2K-Epoxy-Amin Systeme für den Korrosionsschutz, 27. Leverkusener Lacktreff der GDCh-Fachgruppe Lackchemie, 11th September 2014, Leverkusen

[102] M. Koch, Formulating novel aqueous epoxy resin systems for metal primer applications, 6th Nuremberg Congress, April 2001

[103] R. Grasböck, M. Geisberger, Zinkfrei und wässrig – Moderne wässrige EP-Dispersionen und Aminhärter ergeben leistungsstarke Grundierungen, Farbe & Lack 113 (2007) 36–41

[104] M. Gerlitz, A. Krotzek, Breiter einsetzbar – Lösemittelfreie wasserverdünnbare 2K-Epoxy-Primer für den schweren Korrosionsschutz, Farbe u. Lack 114 (2008) 34–37

[105] F. Heine, A new generation of waterborne anticorrosion epoxy resins, Heubach Pigment Symposium 2006, Goslar, Germany 12 to 13. Oct. 2006

[106] M. Gerlitz, Advances in Waterborne 2K Epoxy/Amine Primers for Heavy Duty Applications, 7th Nuremberg Congress (Vincentz, FPL, PRA), 7 to 8 April, 2003, p. 9–19

[107] M. Zirngast, G. Hobisch, F. Lunzer, R. Sanchez-Morillo, Is your waterborne 2K epoxy formulation giving you a headache? Here is the cure. American Coatings Conference, Atlanta USA, 7 to 9 April 2014

[108] H. Kittel, Lehrbuch der Lacke und Beschichtungen, Vol. 2, 2nd Ed., S. Hirzel Verlag Stuttgart Leipzig 1998, p. 299

[109] U. Christ, Untersuchungen zur Haftung von organischen Beschichtungen auf Aluminium und deren Beeinflussung durch Wasser. Dissertation Universität Stuttgart, 1988

[110] P. A. Sørensen, S. Kiil, K. Dam-Johansen, Anticorrosive coatings: a review, J. Coat. Technol. Res., 6 (2) 135–176, 2009, p. 150

[111] ISO 12944-5:2007, Paints and varnishes – Corrosion protection of steel structures by protective paint systems - Part 5: Protective paint systems. International Organization for Standardization, Genève, Switzerland 2007

[112] G. Binder, Examination of Accelerated Laboratory Tests for Corrosion Protection. Protective Coatings Europe (PCE), November 2003, p. 8–16

[113] O. Heins, T. Krebs, M. Baumann, G. Binder, Korrosionsschutz von Offshore-Windenergieanlagen – Einteilung, Normung und praktische Erfahrungen, p. 493–518. HTG-Kongress 2011, 7 to 10. September 2011, Würzburg. Hafentechnische Gesellschaft e.V. Hamburg

[114] Rules and Standards of Federal Waterways Engineering and Research Institute (BAW) Richtlinie für die Prüfung von Beschichtungssystemen für den Korrosionsschutz im Stahlwasserbau (RPB), Bundesanstalt für Wasserbau (BAW) Karlsruhe, January 2011

[115] K. Mühlberg, Surface-Tolerant Coatings – Some Experiences, J. Prot. Coat. Lin. Prot. Coat. Europe Technology Publ. 2001, p. 13ff, 18

[116] J. Sander et al, Anticorrosive Coatings, Fundamentals and New Concepts. Vincentz Network, Hannover, 2010, p. 155–156

[117] DIN 55633:2009, Paints and varnishes – Corrosion protection of steel structures by powder coatings. Beschichtungsstoffe – Korrosionsschutz von Stahlbauten durch Pulver-Beschichtungssysteme – Bewertung der Pulver-Beschichtungssysteme und Ausführung der Beschichtung, Beuth Verlag

[118] Additional Technical Terms of Contract and Guidelines for Civil Engineering Works (ZTV-ING), Zusätzliche Technische Vertragsbedingungen und Richtlinien für Ingenieurbauten ZTV-ING. Bundesanstalt für Straßenwesen, BASt, Stand 11th Dec. 2013

[119] Communication from the German Committee for standardization (DIN) at the Deutsches Institut für Normung DIN-Normenausschuss Beschichtungsstoffe und Beschichtungen (NAB), Subcommittee for Corrosion protective paint systems and testing 2014-06-18

[120] Korrosionsschutz von Stahlbauten durch Beschichtungssysteme. Leistungsbereich DIN EN ISO 12944, Verband der deutschen Lack- und Druckfarbenindustrie e.V. Frankfurt und Bundesverband Korrosionsschutz e.V. Köln, 2010

[121] P. A. Sørensen, S. Kiil, K. Dam-Johansen, Anticorrosive coatings: a review, J. Coat. Technol. Res., 6 (2) 135–176, 2009

[122] DIN-Taschenbuch 286 Korrosionsschutz durch Beschichtungen und Überzüge, Vol. 1 to 4, Beuth-Verlag Berlin, 2012

[123] Additional Technical Terms of Contract and Guidelines for Civil Engineering Works (ZTV-ING), Part 4 Steel Structures, Composite Steel Structures, Section 3 Corrosion Protection of Steel Structures, as of 12/2012 (ZTV-ING 4-3), Stand 11.12.2013

[124] The additional technical terms – hydraulic engineering (ZTV-W) for corrosion protection within hydraulic steelwork (performance range 218) German Ministry of Transport and Digital Infrastructure, The Central Waterways Engineering Library (VZB)of the Federal Waterways Engineering and Research Institute; Drucksachenstelle des WSV beim BSH, Hamburg, 2009

[125] Federal Highway Research Institute 2 2015 - Bundesanstalt für Straßenwesen, BASt, Homepage http://www.bast.de/EN/Structural_Technology/Subjects/e-korrosionsschutz/e-korrosionsschutz.html?nn=604764

[126] M. Schröder, Den Regelfall regeln – ZTV-ING, Bundesverband Korrosionsschutz e.V. (BVK) Jahrestagung 10. to 11. March 2010 Köln

[127] Technical Terms of Delivery and Technical Test Regulations for Coating Materials for the Corrosion Protection of Steel Structures, TL/TP-KOR 2002 respectively TL/TP-ING in preparation: personal communication from Malgorzata Schröder, BASt, 2014

[128] BASt – Compilation of certified coating materials complying with the TL/TP-KOR Steel Structures guidelines maintained by the BASt, Stand 2014

[129] Personal communication from Malgorzata Schröder, BASt, 2014

[130] BAW Guideline for the Testing of Coating Systems for the Corrosion Protection of Hydraulic Steel Structures (RPB), Bundesanstalt für Wasserbau (BAW), Karlsruhe 2011

[131] Merkblatt 405, Korrosionsschutz von Stahlkonstruktionen durch Beschichtungssysteme, Stahl-Informations-Zentrum Düsseldorf, 2000

[132] Richtlinie Korrosionsschutz von Stahlbauten in atmosphärischen Umgebungsbedingungen durch Beschichtungssysteme, Deutscher Stahlbauverband DSTV, Institut für Stahlbau Leipzig GmbH, 1999

[133] In one brochure: Korrosionsschutz durch Duplex-Systeme (Feuerverzinken und Beschichten), Institut Feuerverzinken GmbH Düsseldorf, 2012; pp. 15: Verbände-Richtlinie Korrosionsschutz von Stahlbauten Duplex-Systeme, gemeinsam herausgegeben vom BVK, VdL, DSTV und Industrieverband Feuerverzinken, Düsseldorf, 2000

[134] Quality and Inspection Regulations from the Association for High Quality Corrosion Protection of Valves and Fittings with Powder Coating (GSK), Gütesicherung RAL-GZ 662; Gütegemeinschaft Schwerer Korrosionsschutz von Armaturen und Formstücken durch Pulverbeschichtung e.V. (GSK), Nuremberg, January 2008

[135] International Quality Regulations for the coating of aluminium building components and industrial piecework coating of steel components. GSB AL 631, Steel and galvanized steel GSB ST 663, GSB International e. V. Düsseldorf , May 2013

[136] IQC 654 International Quality Regulation for industrial piecework coating. IQC International; Düsseldorf, November 2012

[137] P. Preuss, Roman Seawater Concrete Holds the Secret to Cutting Carbon Emissions, Berkeley, 2013

[138] V. Smil, Vaulting space – The construction of vaulted roofs through the ages, Nature Volume 408, p. 910, 21 to 28 December 2000

[139] E. Wayman, The Secrets of Ancient Rome's Buildings, 16-11-2011, Internet über: http://www.smithsonianmag.com/history/the-secrets-of-ancient-romes-buildings-234992/?page=1, vom 07 07 2014

[140] R. Boye, S. Foos, J. Hedebratt, G. Plizzari, S. Tattoni, Section 1. Planning and design of industrial floorings, p. 89–134, in TAE Industrieböden, 07/Industrial Floors, 07, Esslingen, 2007

[141] R. Nixon, The fundamentals of cleaning and coating concrete, Pittsburg, USA: The Society for Protective Coatings, 2001

[142] E. Nawy, Fundamentals of high performance concrete, 2nd Ed., New York, USA: John Wiley & Sons, Inc., 2001

[143] N. Williamson, Traning course: Concrete industrial floors: design, detailing and construction, Birmingham: Aston University, 25th April 2006

[144] K. Horie, H. Hiura, M. Sawada, I. Mita and H. Kambe, Calorimetric investigation of polymerization reactions. III. Curing reaction of epoxides with amines, Journal of Polymer Science, Part A1, Vol. 8, pp. 1357–1372, 1970

[145] G. Wisanrakkit and J. Gillham, The glass transition temperature as an index of chemical conversion for a high-T_g amine/epoxy system, Journal of Applied Polymer Science, Volume 41, pp. 2885–2929, 1990

[146] R. Venditti and J. Gillham, A relationship between glass transition temperature and fractional conversion for thermosetting systems, Journal of Applied Polymer Science, Vol. 64, pp. 3–14, 1997

[147] R. Rasing, Unpublished data, Allentown: Copyright Air Products and Chemicals, Inc. Unpublished data, 2014

[148] F. Walker and M. Cook, Two component waterborne epoxy coatings, in ACS Symposium – Technology for waterborne coatings, Series 663, 1997

[149] H.-C. Kuehne, M. Maultzsch and G. Mellmann, Einfluesse auf die Abreissfestigkeit von Betonuntergruenden, in TAE Industrieboeden '07/Industrial Floors '07, Esslingen, 2007

[150] F. Goodwin, Six key points you should know about concrete surface preparation before coating application, JPCL, No. 7, pp. 45–48, 2012

[151] R. Higgins, http://www.carpetinspector.com/members/secret_enemy_of_flooring_install.htm, Internet on 7–4–2014: http://www.carpetinspector.com/members/secret_enemy_of_flooring_install.htm

[152] R. Higgens, Dew Point and Condensation problems with Flooring Installations, 5–29–2013 Internet on 7–4–2014: http://www.pbawichita.com/blog/4/dew-point-and-condensation-problems-flooring-installations-0

[153] S. Horstmeyer, http://www.shorstmeyer.com/wxfaqs/humidity/humidity.html, 2008, Internet on 7–4–2014: http://www.shorstmeyer.com/wxfaqs/humidity/humidity.html

[154] CCAA, Datasheet: Moisture in concrete and moisture-sensitive finishes and coatings, 4/2007. Internet on 7–5–2014: http://www.concrete.net.au/publications/pdf/Moisture.pdf

[155] T. Lohmann, Vermeiden von Blasenbildung beim beschichten poroeser Untergründe, 275–283, in TAE Industrieboeden '07/Industrial Floors '07, Esslingen, 2007

[156] G. Pleyers, Method for sealing porous building materials and building components. Patent EP 110273B1, 23 June 1999

[157] C. Grave, I. McEwan and R. Pethrick, Influence of stoichiometric ratio on water absorption in epoxy resins, J. Applied Polymer Science, vol. 69, pp. 2369–2376, 1998

[158] R. Gaul, Moisture-Caused Coating Failures: Facts and Fiction, Concrete Repair Digest, no. February–March, 1997

[159] 3rd E. Guideline for Planning and Executing Waterproofing on Building Elements, Deutsche Bauchemie, Frankfurt a.M., May 2010

[160] F. Pfaff and F. Gelfant, Osmotic blistering of epoxy coatings on concrete, JPCL, no. 12, pp. 52–64, 1997

[161] L. Wolff, M. Raupach and K. Hailu, Mechanismen der Blasenbildung bei Reaktionsharzbeschichtungen auf Beton, 35–46, in TAE Industrieböden '07/Industrial Floors '07, Esslingen, 2007

[162] G. Rheinwald, Osmoseblasen an einer rissüberbrückenden EP-Beschichtung, 255–259, in TAE Industrieböden '07/Industrial Floors '07, Esslingen, 2007

[163] R. Cain, Controlling moisture migration with advances in polymer technology, in Thermoset Technology Division, Cincinnati, USA, 13–15 October 2002

[164] M. Lohe, M. Cook and A. Klippstein, Three-dimensional epoxy binder structures for water damp permeable and breathable coating and flooring systems, Macromol. Symp., Vol. 187, pp. 493–502, 2002

[165] R. Frick, Rückwärtige Durchfeuchtung erdberührender Betonplatten, VBK Bautenschutz, March 2006

[166] W. Ashcroft and M. Scargill, Evaluation of the susceptibility of polyamine curing agents to carbonation/waterspotting at 5–25 degrees Celsius (Technical bulletin #115/C), Air Products and Chemicals, Inc., Allentown, 1998

[167] J. Bell, J. Reffner and S. Petrie, Amine-cured epoxy resins: adhesion loss due to reaction with air, Journal of Applied Polymer Science, pp. 1095–1102, 1977

[168] P. Lucas, P. Clark, R. Haney and M. Kittek, Investigation of water spot and blush resistance of epoxy industrial floors, Journal of Protective Coatings and Linings, pp. 20–27, 1998

[169] P. Flory, Journal of American Chemical Society, Vol. 63, pp. 3083–3090, 1941

[170] N. D. Bruyne, Modern plastics, Vol. 27, p. 9, 1950

[171] B. Hills, J. Oil and Colour Chemists Assoc., Vol. 45, pp. 251–260, 1962

[172] R. Rasing and G. Vedage, A breath of fresh air in emission free flooring: Novel epoxy curing agent technology introduces improved formulation latitude, high performance and clean indoor air, in European Coatings Congress, Nuremberg, 2011

[173] R. Rasing and J. Bender, Beyond Green: A toolbox for uncompromised industrial flooring, Polymer Paint Colour Journal, No. 5, pp. 24–28, 2013

[174] R. Young and P. Lovell, Introduction to polymers, 2nd Ed., p. 419–420, London: Chapman & Hall, 1991

[175] P. Lucas, W. Starner, S. Musselman and J. Reader, Maximizing toughness in epoxy polymers through the use of acrylate-functional urethanes, in SPI-ERF conference, Boston, USA, 1994

[176] D. Schedlitzki and W. Slama, Measuring crack-bridging properties in linings for concrete in chemical exposures, Journal of Protective Coatings and Linings, No. 2, pp. 25–37, 2000

[177] Conductive coatings for industrial floors, 1st Ed., Deutsche Bauchemie e.V., Frankfurt a.M., June 2003

[178] G. Hall, Controlling electrostatic discharge with static control floor coatings, Journal of Protective Coatings & Linings, No. 8, pp. 56–63, 1996

[179] F. Hazen, Repair and resurfacing of paper mill bleach plant floors and secondary containment, JPCL, No. 8, pp. 60–67, 1993

[180] C. Stanley, Chemical resistant epoxies for secondary containment, JPCL, No. 12, pp. 17–22, 1995

[181] M. McGovern, Protecting floors from chemical attack, Concrete Construction, No. 10, 1998

[182] J. Ratliff, Protecting concrete floors in the food and beverage industry: a comparative look at floor systems, JPCL, No. 4, pp. 55–65, 2007

[183] Council Directive 89/106/EEC of December 21 1988 on the approximation of laws, regulations and administrative provisions of the Member States relating to construction products, OJ L 40, 11 February 1989; often referred to as Construction Products Directi, European Committee, Brussels, 1989

[184] J. R. C. E. I. European Commission, European Collaborative Action Indoor Air Quality and Its Impact On Man, Evaluation of VOC emissions from building products; solid flooring materials (Report No. 18), EUR17334EN, Brussels, 1997

[185] J. R. C. I. f. H. a. C. P. European Commission, European Collaborative Action Urban Air, Indoor Environment and Human Exposure, Harmonisation of indoor material emissions labeling systems in the EU; inventory of existing schemes (Report No. 24), EUR21891EN, Brussels, 2005

[186] D. I. f. B. (DIBt), Ausschuss zur Gesundheitlichen Bewertung von Bauprodukten (AgBB), Bewertungsschema für VOC aus Bauprodukten, Deutsche Bauchemie, Frankfurt a.M., 2008

[187] A. Crowe, A. Toullec, R. Oppl, S. Mulholland and R. Augustin, Seminar – VOC emissions testing in Europe, Birmingham, UK: Eurofins, 5 February 2013

[188] F. Beyeler, N. Beglinger and U. Roder, Minergie: The Swiss Sustainable Building Standard, MIT Press` Innovations, vol. 4, No. 4, pp. 241–244, 2009

[189] U. G. B. Council, http://www.usgbc.org/leed#rating, 2014. Internet on 7–10–2014: http://www.usgbc.org/leed#rating

[190] S. Monaghan and R. Rasing, Waterborne epoxies – a practical, economic solution to low emission industrial floorings, in European Coatings Congress, Nuremberg, 2009

[191] R. Rasing, Copyright Air Products and Chemicals, Inc., Value chain of applied flooring technology, 2014

[192] R. Young and P. Lovell, Introduction to polymers, 2nd Ed., London: Chapman & Hall, 1991

[193] C. Van Ginderachter and B. Parmentier, Influence of the boundary conditions and reinforcement on differential shrinkage of concrete slabs, in TAE Industrieböden '07/Industrial Floors '07, Esslingen, 2007

[194] N. Yuasa, T. Sasaki, I. Matsui and Y. Kasai, Effect of porosity and moisture content of concrete slab on osmotic bliser of polymer, 269–274, in TAE Industrieböden '07/Industrial Floors '07, Esslingen, 2007

[195] H. Kittel, Lehrbuch der Lacke und Beschichtungen, Vol. 2, 2nd Ed., S. Hirzel, Stuttgart, 1998, p. 293–296

[196] D. Stoye, W. Freitag, Lackharze. Chemie, Eigenschaften und Anwendungen. Carl Hanser Verlag München, Wien, 1996, p. 275

[197] B. Müller, U. Poth, Coatings Formulation, 2nd Edition, Vincentz Network, Hannover, 2011, p. 262–272

[198] J. Pietschmann, Industrielle Pulverbeschichtung, 3rd Ed., Vieweg + Teubner, GWV Fachverlage, Wiesbaden 2010, p. 5–24

[199] T.A. Misev, Powder Coatings, Chemistry and Technology, John Wiley & Sons Ltd. Chichester, England, 1991, p. 117ff.

[200] Alzchem-Broschüre "Dyhard" – Curing Agents for Powder Coatings 04/2014

[201] DE102006056311A1 05.06.2008: Verwendung von Guanidin-Derivaten als Beschleuniger bei der Härtung von Epoxidharzsystemen

[202] WIPO Patent Application WO/2004/106402: Use of urea derivatives as accelerators for epoxy resins Publication date december 09, 2004

[203] Huntsman Advanced Materials Solid building blocks, Quarter 1, 2014, p. 3

[204] "Aradur" 2844 Data Sheet, Huntsman Advanced Materials, Basel 04/2008

[205] Alzchem-Product Information "DYHARD" – Curing Agents – DYHARD PMC, 01–2014

[206] Technical Data Sheet PE1204A, FreiLacke, Bräunlingen, 02/2012

[207] H. Kittel, Lehrbuch der Lacke und Beschichtungen, Vol. 2, 2nd Ed., S. Hirzel, Stuttgart, 1998, p. 307

[208] Z. W. Wicks, F. N. Jones, S. P. Pappas, Organic Coatings: Science and Technology, Volume I: Film Formation, Components, and Appearance. J.Wiley & Sons, New York, 1992, p. 177

[209] Technical Data Sheet Huntsman Advanced Materials "ARADUR" 9690-1 Hardener, The Woodlands, Texas, June 2012

[210] T. A. Misev, Powder Coatings, Chemistry and Technology, John Wiley & Sons Ltd. Chichester, England, 1991, p. 141
[211] D. Stoye, Paints, Coatings and Solvents, VCH Verlagsgesellschaft Weinheim, 1993, p. 72
[212] H. Kittel, Lehrbuch der Lacke und Beschichtungen, Vol. 2, 2nd Ed., S. Hirzel, Stuttgart, 1998, p. 161 ff.
[213] DSM Product Overview Powder Coating Resins Europe BV Zwolle, 2011
[214] Allnex Powder Coating Resins, Anderlecht – Brussels, Belgium 2013
[215] P. Mischke, Filmformation, Vincentz Network, Hanover, 2010
[216] Pulverbeschichtung von MDF, EFD-Info No. 507, FreiLacke, Bräunlingen, 11/2012
[217] M. Badila, C. Jocham, W. Zhang, T. Schmidt, G. Wuzella, U. Müller, A. Kandelbauer, Powder coating of veneered particle board surfaces by hot pressing. Progress in Organic Coatings 77 (2014) 1547–1553
[218] G. Wuzella, A. Kandelbauer, A. R. Mahendran, U. Müller, A. Teischinger, Influence of thermo-analytical and rheological properties of an epoxy powder coating resin on the quality of coatings on medium density fibreboards (MDF) using in-mould technology. Progress in Organic Coatings 77 (2014) 1539–1546
[219] J. Keller, Energieeffiziente Pulverlacke, Metalloberfläche mo 63 (2009) 4, 20–21
[220] Araldite PT 910 Technical Data Sheet, Huntsman Advanced Materials, Basel, Oct. 2012
[221] "CRYLCOAT" 4540-0, Technical Data Sheet, Allnex Powder Coating Resins, August 2013
[222] P. Gottis, Polyester/Araldite PT 910 Powder Coating Formulations, Huntsman Advanced Materials, 15.10.2004
[223] P. Gottis, Araldite PT 910 Guiding Notes for optimal dispersion under laboratory conditions, Huntsman Advanced Materials, 09.11.2004
[224] GSB – Internationale Qualitätsrichtlinien für Beschichtung von Bauteilen aus Aluminium GSB AL 63, Stahl und feuerverzinktem Stahl GSB ST 663, Ausgabe Mai 2013, GSB International e. V. Düsseldorf
[225] Qualicoat Specifications for a quality label for liquid and powder organic coatings on aluminium for architectural applications, 13th Ed. 2012, Zürich
[226] M. Hilt, U. Christ, Solutions without solvents, trends, technologies and perspectives in the replacement of wet paints, Europ. Coat. Journal 12, 2010, p. 82–87
[227] "CRYLCOAT" 4540-0, Starting Formulation Polyester, Allnex Powder Coating Resins, 2013
[228] S. Lu, Innovations in GMA acrylic powder coatings, Powder Coating February 2013, p. 20–24
[229] U. Christ, Lackierung von Leichtmetallrädern – Schöner Schutz, JOT 4 (2008), p. 2–4
[230] Anderson Development company – Technical Information Almatex AP4411 GMA Powder Clearcoat, SL 130613, 2013
[231] http://www.efsa.europa.eu/de/topics/topic/bisphenol.htm

4 Trends and outlook

Ulrich Christ

Because of outstanding properties associated with the chemistry of epoxy resins, such as adhesion, wet adhesion, mechanical strength, corrosion protection, chemical, mechanical, thermal, and hydrothermal stability, minimal shrinkage and electrical insulation, the range of applications of epoxy resins is very large and their economic impact is very prominent.

All participants of the process chain, from producers of raw materials through the manufacturers of coating materials to the end user, have a great interest in the advancement of epoxy resins with respect to the requirements of, among other things, coating materials for corrosion protection, for surface coatings, concrete protection and flooring as well as the rapidly growing field of matrix resins for composites.

4.1 Legal requirements related to health, safety and environmental protection

Important driving forces for new developments in the field of coating materials for both corrosion and surface protection (including flooring) are the legal regulations regarding the health protection, occupational safety and the limitation of **v**olatile **o**rganic **c**ompounds (VOC) and **s**emi **v**olatile **o**rganic **c**ompounds (SVOC), the **r**egistration, **e**valuation, **a**uthorization and restriction of **ch**emicals (REACH regulation), the registration requirement of substances in Annex XIV of REACH as well as the list of SVHCs (**s**ubstances of **v**ery **h**igh **c**oncern). A few years ago, the committee for health-related evaluation of construction products (**A**usschuss für **g**esundheitliche **B**ewertung von **B**auprodukten (AgBB) [1], was been founded, which evaluates the health risk of VOC emissions and supports the development of particularly low-emission products for the implementation in construction products for closed rooms in which people reside longer.

4.2 New product developments

The manufacturers of raw materials and paints, users or the using industries and their representative organizations are striving, on the one hand, to adjust their objectives and the need for regulation to the state of knowledge regarding the effect of substances on the environment and the human beings and, on the other hand, to find a way to implement the aforementioned regulations or restrictions in the product development and in technical processes. Therefore, there is a trend towards developments of 100 % reactive or low-emission products, as demonstrated by the following examples of enhancements of 2pack-epoxy resins and hardeners, 1pack-epoxy resins for corrosion protective coatings, 2pack-epoxy resins and hardeners for topcoats, as well as replacement products for bisphenol A (BPA) in can coatings for the food industry.

Further development of 2pack-epoxy binders and curing agents is needed for high-solids and solvent-free systems, especially for the reduction of emissions and the sensitization potential of reactive diluents and curing agents in the processing state of the therewith prepared coating materials. Even with epoxy resins and curing agents for waterbased coatings, there is still need for further development in terms of VOC and drying at low temperatures and high humidity. Since the lowest drying temperature is limited by the minimum film formation temperature (MFT) of the epoxy resin dispersion, the reduction of VOCs and the lowering of MFT are partially contradictory demands.

Many application characteristics, such as a possibly long application period for the processing of the coating material, oppose the requirements for lower VOCs and for products with the lowest possible sensitizing potential. Therefore, in most of the cases, a compromise between the technical performance of the products and the requirements for these products and their content of constituents, possibly harmful to health and environment, has to be found. Solvent-free and yet readily processable products require for technical functionality a certain proportion of low molecular weight substances, which should possibly not reveal any sensitizing effect.

With respect to the aforementioned aspects, new, accelerated and 100 % reactive PAA hardener systems for 2pack-epoxy coating materials show a very interesting property profile [2]: the hardener systems are free of SVHC, free of benzyl alcohol, free of alkyl phenols and free of plasticizers. They are also suitable for AgBB related applications and facilitate high-solid 2pack-epoxy coating materials with VOC levels reduced by 15 % as compared to standard high solid systems. Thus producible coatings show the same, on critical substrates even a better corrosion protection and better adhesion than standard solventbased systems. However, this new class of products for obtaining coatings with high elasticity currently reveals a PVC (pigment volume concentration) limit at a maximum of 25 to 30 % [2].

4.2.1 Epoxy resins – applicable in future also for topcoats

Up to now, 2pack-epoxy/amine-based coatings cannot be used as topcoats due to UV degradation of aromatic structures with the consequence of strong reduction of gloss, chalking and the strong yellowing when exposed to weathering. Recently, as a solution for the reduction of the UV-initiated degradation, non-aromatic epoxy resins with improved gloss retention, improved chalking resistance and less yellowing have become available [2]. The new UV-stable 2pack-epoxy coating system is based on an aliphatic, polyfunctional trimethylolpropane (TMP)-triglycidylether or a cycloaliphatic epoxy compound based on hexahydro-phthalic anhydride (PSA) diglycidyl ether and a new 100 % reactive amine hardener based on aliphatic/cycloaliphatic building blocks, free of aromatics. 2pack-epoxy coating materials on this binder/hardener basis can be formulated with zero VOC and suitability for AgBB relevant applications. They show a gloss retention level under weathering conditions, which is suitable for industrial topcoats. The yellowing is indeed greater than for 2pack-polyurethane (PUR) topcoats, which are cured with aliphatic isocyanates, but it is acceptable for some industrial applications, such as agricultural and construction machinery (ACE) as well as for rail vehicles. The mechanical properties of this topcoat still require optimization, because cycloaliphatic epoxy compounds tend to brittleness, which allows applications in the flooring area. Epoxy topcoats having good mechanical properties can be achieved by combination of multifunctional aliphatic epoxy resins with flexibilized non-aromatic reactive diluents [2]. Such new epoxy resin and hardener building blocks open up a path to product rationalization in epoxy primer and topcoat coating structures: for the primer and the top coating material a common hardener could be used.

4.2.2 New waterbased 1pack-epoxy technology for high duty corrosion protection systems

Up to now, water-dilutable 1pack-coating systems usually only meet the requirements for corrosion protection in the corrosivity categories C1 to C3 (M), corresponding to a resistance in the salt spray test up to about 250 h of loading.

Water-dilutable or high solid 2pack-epoxy systems are used for the corrosion protection with higher corrosive requirements. Besides the excellent corrosion protection of water-dilutable 2pack-epoxy systems, also disadvantages of these products are discussed, in particular the handling and the highly temperature-dependent curing reaction. During the processing and curing the minimum film formaton temperature (MFT) is particularly important, which increases with the start of curing reaction (building of crosslinking network) and augmenting glass transition temperature (T_g). If during the processing under field conditions the substrate temperature decreases to MFT, the film formation stops. Furthermore, 2pack-coating systems in general need at least one week of curing, until the final properties are achieved.

These two main aspects motivate the development of 1pack-binders, which perform similarly to 2pack-systems.

Recently, the development of a physically drying 1pack-epoxy resin technology, with a branched structure similar to 2pack-epoxy crosslinking chemistry and a high level of corrosion protection, was reported [3]. In this case, using the knowledge about cataphoresis coatings, the synthesis of the binder was carried out using epoxy amino resins, which have good anti-corrosion properties. Unlike in conventional 1pack-acrylic dispersions, the molecular weight was not built-up by linear chain growth but by high degree of branching, similar to the 2pack-epoxy crosslinking. The resulting new, cationically stabilized dispersion of high molecular weight epoxy-amine adducts cures entirely physically and reaches the final properties, such as water resistance, corrosion protection and chemical resistance, very fast. Thereby, the MFT of the binder remains constant at 5 °C, enabling the application of the product for corrosion protection under real conditions. This new 1pack-epoxy technology facilitates primer formulations with VOC levels of 73 g/l; primers produced therefrom, followed up by applying 2pack-PUR hydro topcoats, achieve a corrosion protection corresponding to the corrosivity category C4 and very good water resistance. As shown by studies of coatings with this binder basis on various substrates, previous 1pack-coating systems are surpassed in terms of corrosion protection and water resistance even after drying at temperatures below 10 °C. This new epoxy technology is attributed the potential to close the gap between the high efficiency of 2pack-epoxy systems and the standard 1pack-systems for primer and one-layer systems for corrosion protection [3].

4.2.3 Improving the corrosion protection of 2pack-epoxy coatings by active anti-corrosion and barrier pigments

Another important topic is the improvement of corrosion protection of 2pack-epoxy coatings by active and barrier pigments, fillers and inhibitors. There is a continuing development trend of novel and optimized active pigments for compatibility with certain binder systems [4, 5]. The addition of active anti-corrosion pigments results in a strengthening of the inherent barrier and anti-corrosive effect of epoxy resins in coating formulations. Thus they can also help to achieve a certain level of corrosion protection by a lower total layer thickness or to extend the maintenance and renewal intervals for anti-corrosion coatings on the object.

Active and barrier pigments containing nanoparticles are also a field of research in connection with epoxy systems in terms of their contribution to improving the corrosion protection. In a research project, the corrosion protection effect of nanoscale active pigments, obtained from the grinding of standard products, was studied and compared to standard pigments in a 2pack-epoxy primer, applied on blasted steel [6]. It was shown that the nanoscale anti-corrosive pigments are dispersed more homogeneously in the priming coat and better penetrate into micro areas of the jagged substrate surface than the standard materials. The better performance of nanoscale pigments also manifests itself in a higher electrochemical homogeneity of the interface and a less active corrosion near an artificially placed defect. Results from electrochemical short-term tests, such as electrochemical impedance spectroscopy and scanning Kelvin probe, correlated well with results of the corrosion tests using the salt spray test and weathering on Helgoland (outdoor weathering station on an island in the North Sea). They show that 2pack-epoxy priming coats with nanoscale anti-corrosive pigments on steel, galvanized steel and aluminium achieve a better corrosion protection than with standard active pigments [6].

The influence of nanoscale barrier fillers on the corrosion protection and other coating properties is also the subject of research. Investigations on 2pack-epoxy corrosion protection coatings with epoxy layer silicate nanocomposites, consisting of BPA-based solid epoxy resin solution, a polyamidoamine (PAA) hardener and a natural montmorillonite, modified with a quaternary ammonium salt, reveal the following characteristics: the corrosion protection was the largest at the lowest concentration of the layer silicate composite (at 1 %) [7]. This correlates with the highest found concentration of exfoliated nanoscale layer platelets and the biggest barrier effect, respectively. The content of exfoliated layered silicate platelets decreases with increasing concentration of layer silicate. It was also shown that, at low concentrations of the layer silicate, the glass transition temperature T_g and the storage module in both the glass and in the rubber elastic state increase and that the thermal stability of the coatings is improved.

In the course of spreading renewable energy generation, 2pack-epoxy coatings are of great importance. In a recent study of the corrosion protection of offshore wind turbines in the underwater zone, the low tide zone and the splash zone, duplex systems, consisting of a zinc coating or an aluminium metal spray coating on steel and three epoxy intermediate coatings with partial reinforcement particles with a total film thickness of 920 μm, were best-performing [8]. Thereby, different corrosion protection systems with the following target parameters were taken into account: corrosion protection, adhesion of the coating, fouling, flange corrosion and coating behaviour of welded and bolted connections. Further research demand is seen in the corrosion protection of offshore wind turbines, especially at the foundation structures, such as monopiles, and on steels in the seabed as well as in the field of microbially induced corrosion [9].

Another focus of research is on the special chemical, mechanical and thermal stability of epoxy coatings, particularly in terms of the high variety of binder and hardener combinations. An excellent example is the thermal-oxidative stability of epoxy coatings crosslinked with different hardeners that were studied at temperatures from 70 °C to 150 °C by means of FTIR spectroscopy, DSC and sol-gel analysis (SGA). Thermal-oxidative aging of the coating, cured with polyamidoamine (PAA), preferably induced a post-curing while maintaining or increasing the glass transition temperature T_g, whereas polyoxypropyleneamine (POPA) cured 2pack-epoxy coatings are more sensitive under high temperature and prone to chain scission, which is associated with a decrease of the glass transition temperature T_g [10].

4.2.4 Trends in epoxy-based powder coatings

The wide variety of crosslinking chemistry of epoxy resins with hardeners covers the main application fields of powder coatings: high corrosion-resistant primers, the universal application of hybrid powder coatings for decorative purposes, low temperature powder coatings for medium-density fibreboard (MDF) and other thermo-sensitive materials, the glycidyl esters that allow super-durable polyester powder coatings for outdoor applications and the glycidyl methacrylate (GMA) crosslinking agents for highly brilliant automotive clear-coats with good levelling.

Epoxy-based powder coatings will further have a very great economic importance for the functional (corrosion protection) and decorative applications in the future. Powder resin and curing agent technologies, suitable for poly-addition, are a good prerequisite for the production of high-quality, low-defect coatings. With appropriate catalysts, the epoxy powder coating technology enables even more applications in the field of eco-efficient coatings of thermo-sensitive materials such as plastics, among others thermoplastics and thermosets as well as of wood-based materials, as shown by recent developments, e.g. in the field of MDF, veneer and plywood coatings [11]. It has been reported that high-gloss single-layer powder coatings with a very good property profile are achievable. This is likely to be a promising coating material and process technology, whose market development has only just begun.

Even with epoxy-based powder coatings, studies on the improvement of the barrier effect are underway [12]. It turns out that powder coating formulations with organically modified phyllosilicates, based on montmorillonite in 2 to 4 % portions, also increase the glass transition temperature T_g and the thermal stability of powder coatings.

In analogy to liquid coating materials, active anti-corrosion pigments are used in powder coatings to improve the corrosion protection. An epoxy polyester hybrid powder coating with an active pigment, based on modified zinc calcium phosphate, shows after 1,000 hours of salt spray test no under-creepage in contrast to a sample without anti-corrosive pigment, which reveals significant under-creepage [13].

There are new developments in acrylate powder coating materials ongoing, which exhibit improved compatibility with polyester powder coatings and can be used as clearcoats with very good levelling and as pigmented coatings. In addition, there is also progress in curing acrylate powder coatings at lower temperatures (for example at 145 °C) [14, 15].

4.3 Potential replacement of BPA in the can coatings industry

Michael Dornbusch

Research into the use of epoxy resins in food-grade packaging, especially in the field of can coatings (see Section 3.5.1), has for some years focused primarily on replacing BPA in coatings on account of its endocrinal effect (see Sections 1.1 and 3.5.1) and will continue for several years yet. There are three principal strategies for replacing BPA.

4.3.1 Replacing BPA with derivatives of bisphenol A

From the technological point of view, the obvious way to replace BPA is to use other bisphenols. Currently under discussion [16, 17] are compounds such as bisphenol F or C (see Table 2.4). Their structures suggest that they have no endocrinal effect [17], but they could affect other properties of the resins, such as glass-transition temperature, chemical resistance, and flexibility. It is therefore questionable as to whether any of the known bisphenols could serve as an alternative to BPA. In the literature, bisphenol S (BPS) [18] (see Figure 4.1, left) has been proposed, but is also suspected of having endocrinal effects. By contrast, the monomer (R,S)-1,1-bis(4-hydroxyphenyl)-3,3,5-trimethylcyclohexane (TMC) is not expected to have

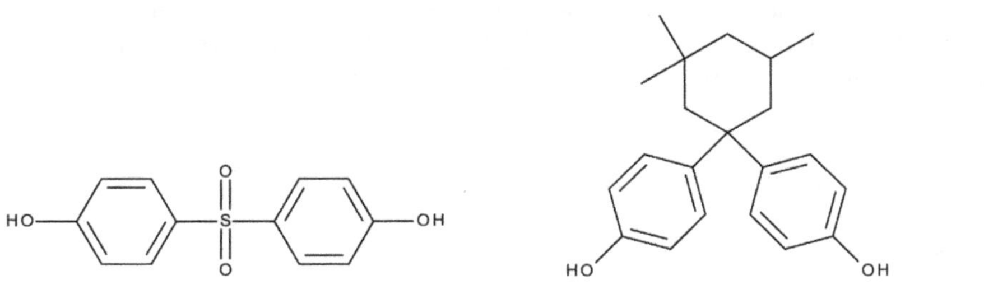

Bisphenol S

(R,S)-1,1-Bis(4-hydroxyphenyl)-3,3,5-trimethylcyclohexane

Figure 4.1: Structures of BPS (left) and TMC (right)

an endocrinal effect, but of course the properties which it imparts to the resultant resin are different from those produced by BPA [19]. The monomer has already been successfully used as an alternative to BPA in polyimides, and has improved some property profiles (see Figure 4.1, right).

4.3.2 Replacing BPA with new epoxy compounds

The second strategy consists in searching for new diols, but limiting the search to candidates which are non-toxic, non-CMR and which have no endocrinal effect. Furthermore, the resultant resins need to have a property profile similar to that of resins containing BPA.

One such diol is 2,2,4,4-tetramethyl-1,3-cyclobutanediol (TMCD), which is already being used in polyesters by Eastman [18] as a replacement for BPA-containing polycarbonate [18, 19]. Given that reaction to the corresponding diglycidyl ether is possible [20] (see Figure 4.2), this diol could be a good candidate for resins in the coatings industry.

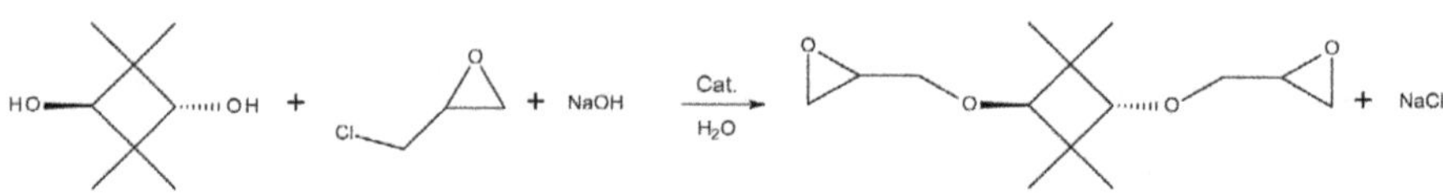

Figure 4.2: Synthesis of the diglycidyl ether of TMCD. The catalyst employed is tetrabutylammonium hydrogensulfate [20]

The third, highly sustainable strategy consists in synthesising epoxy resins from isosorbide (Figure 4.3), which is derived from glucose by hydrogenation to D-sorbitol, followed by acid-catalysed dehydration and is therefore a renewable raw material [21]. Isosorbide has already been used to produce diglycidyl ethers by reaction with epichlorohydrin [19, 22]. The ethers are then cured with different various crosslinking agents to yield different coating systems. The coatings show physical properties similar to those containing BPA, but their chemical resistance, which is crucial for can coating, is substantially lower [19].

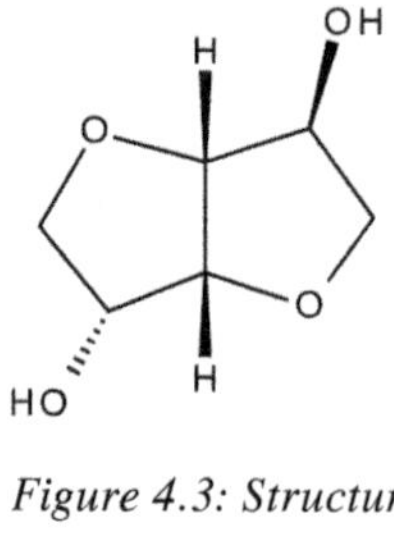

Figure 4.3: Structure of isosorbide

Both diols mentioned above are pure cycloaliphatic systems, lacking the aromatic system that is needed for providing outstanding corrosion protection to the resultant coating. Corrosion protection is the second key property governing use in can coating. The recently patented 2-phenyl-1,3-propanediol [23] (Figure 4.4), which can be transformed into the corresponding diglycidyl ether, contains an aromatic group and possesses a structure that suggests it has no endocrinal effects. The resultant epoxy resins have a lower viscosity than their BPA analogues and the corresponding coatings have a lower T_g; however, the level of corrosion protection and chemical resistance are highly encouraging for use in can coating [23].

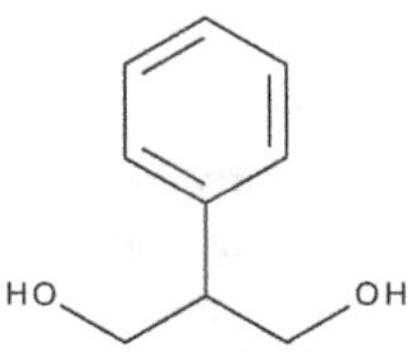

Figure 4.4: 2-Phenyl-1,3-propanediol

A further group of aromatic candidates is composed of 1,3,4-oxadiazoles, which are synthesised by reaction of hydrazine with benzoic acids [19] (see Figure 4.5). The resultant coatings possess improved thermal stability and higher fracture toughness.

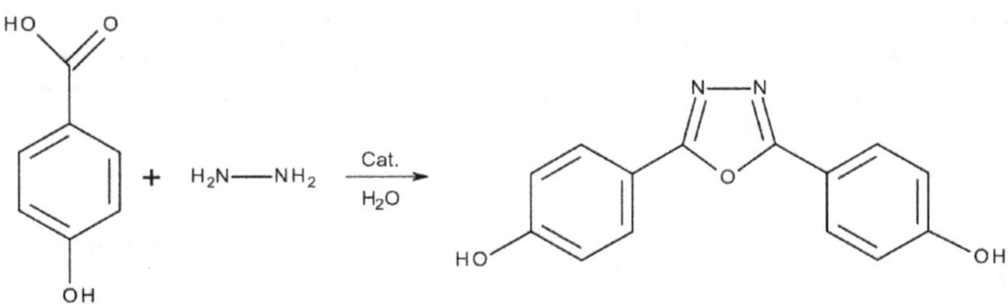

Figure 4.5: Synthesis of 1,3,4-oxadiazoles. The catalyst employed is tetrabutylammonium hydrogensulfate [19]

4.3.3 Replacing BPA with other resin types

Of course, there are numerous commercial developments available which are aimed at replacing epoxy resins containing BPA with other types of resin. Trials have mainly focused on blending different resins [18], e.g. blends of polyesters with polyacrylates or polyesters with polyurethanes, as well as more complex blends of polyacrylates with polyvinyls and polyesters. These trials have failed to generate the requisite property profile and have succeeded merely in lowering the BPA content of the formulations, i.e. the requisite property profile has been obtained by using blends with epoxy resins containing BPA [19].

4.4 Epoxides as building blocks for use of anthropogenic carbon dioxide for chemical syntheses

Ulrich Christ

Epoxides play an entirely new role in the use of anthropogenic carbon dioxide as a sustainable source of raw materials for chemical synthesis. Via catalysed coupling of carbon dioxide to epoxides, customized polyether carbonate polyols become accessible as building blocks for polymers [24]. As a result, the carbon footprint of CO_2-based polyol production can be significantly lowered as compared to the conventional syntheses [25].

4.5 Outlook – a strong growth predicted for epoxy resins

The global epoxy market is expected to grow to € 7.1 billion until 2019, corresponding to an average annual growth of 7.4 %. The increasing demand for epoxy resins in the wind energy and the coating materials especially contributes to this growth [26]. Accompanied by the worldwide growing demand for steel in construction and infrastructure, there is also considerable interest for epoxy-based coating materials for corrosion protection.

Growth through the expansion of lightweight construction and renewable energy generation

An increasingly important significance of epoxy resins is obvious for fibre-reinforced composites, inter alia, in the field of renewable energy generation for the production of wind turbine blades, sports equipment, and for the lightweight technology, as for the aircraft and automobile industries. According to a published report, the market for mass production of high-strength fibre composite components for the lightweight construction is expected to grow with 17 % per year until 2020 in demand of carbon fibre reinforced plastics (CFRP). The main growth drivers for these applications are the automotive, aerospace, wind turbines, and the mechanical and plant engineering [27]. A recent study [28] anticipates an annual increase of this product line by 10 % until 2020. According to this study, the main applications are the aerospace, sports and leisure, wind turbines, and the vehicle construction.

4.6 Literature

[1] AgBB – The Umweltbundesamt (UBA)-Ausschuss zur gesundheitlichen Bewertung von Bauprodukten. UBA Committee for health-related evaluation of building products. http://www.umweltbundesamt.de/en/search/content/AgBB

[2] M. Gerlitz, A. Dureault, A. Napoli, P. Luethi, New 2-layer Epoxy Coating System Achieving Excellent Corrosion Protection and UV-Stability. ETCC Congress 3rd of September 2014, Cologne, Germany

[3] M. Zirngast, R. Feola, F. Lunzer, Air-drying waterborne one-pack coatings for corrosion protection, SSPC Conference 2013, 14th–17th January 2013, San Antonio, Texas USA

[4] S. Bender, M. Babutzka, L. Kirmaier, Korrosionsschutz schnell untersuchen, Farbe u. Lack 120 (5) (2014), p. 22–28

[5] Nubiola Nubirox Technical Guide 1. Ed., Barcelona, January 2009

[6] M. Entenmann, H. Greisiger, R. Maurer, T. Schauer: Corrosion Protection with Nanoscale Anticorrosive Pigments in Coatings, Europ. Coatings Congress, 29th of March 2011, Nuremberg

[7] M. D. Tomic, B. Dunjic, V. Likic, J. Bajat, J. Rogan, J. Djonlagic, The use of nanoclay in preparation of epoxy anticorrosive coatings, Progress in Organic Coatings 77 (2014) p. 518–527

[8] A. W. Momber, P. Plagemann, V. Stenzel, Performance and integrity of protective coating systems for offshore wind power structures after three years under offshore site conditions, Renewable Energy 74 (2015) p. 606–6178

[9] O. Heins, T. Krebs, M. Baumann, G. Binder, Korrosionsschutz von Offshore-Windenergieanlagen – Einteilung, Normung und praktische Erfahrungen, S. 493–518, HTG-Kongress 2011, 7th –10th of September 2011, Würzburg. Hafentechnische Gesellschaft e.V. Hamburg

[10] Y. Zahra, F. Djouani, B. Fayolle, M. Kuntz, J. Verdu, Thermo-oxidative aging of epoxy coating systems. Progress in Organic Coatings 77 (2014) p. 380–387

[11] G. Wuzella, A. Kandelbauer, A. R. Mahendran, U. Müller, A. Teischinger, Influence of thermo-analytical and rheological properties of an epoxy powder coating resin on the quality of coatings on medium density fibreboards (MDF) using in-mould technology, Progress in Organic Coatings 77 (2014) p. 1539–1546

[12] D. Piazza, N. P. Lorandi, E. S. Rieder, L. C. Scienza, and A. J. Zattera, Epoxy-Montmorillonite Nanocomposites Applied to Powder Coatings, International Polymer Processing: Vol. 26, (2011) No. 5, p. 478–483

[13] F. Deflorian, S. Rossi, M. Fedel, L.G. Ecco, R. Paganica, M. Bastorolo, Study of the effect of corrosion inhibitors on powder coatings applied on steel, Progress in Organic Coatings 77 (2014) p. 2133–2139

[14] S. Lu, Innovations in GMA acrylic powder coatings, Powder Coating February 2013, p. 20–24

[15] Anderson Development Company – Technical Information Almatex AP4411 GMA Powder Clearcoat, SL 130613, 2013

[16] Preferred Alternative Technology to BPA-based Epoxy Can Coatings, SpecialChem, April 1, 2014

[17] WO2012/091701 A1

[18] M. M. Bomgardner, No easy fix for food can coatings, C&EN Northeast news bureau, February 11, 2013

[19] A. M. Nelson, T. E. Long, Polym. Int. 2012, 61, p. 1485–1491

[20] WO2012/149340A1

[21] F. Fava, P. Canepa, INCA, Venezia, F. W. Lichtenthaler, Industrial chemicals from Carbohydrate feedstocks: current status and challaenges ahead, 2008, p. 230–253

[22] US 2008/0009599A1

[23] US 2014/0128503A1

[24] J. Langanke, A. Wolf, J. Hofmann, K. Böhm, M.A. Subhani, T.E. Müller, W. Leitner, C. Gürtler, Carbon dioxide (CO_2) as sustainable feedstock for polyurethane production, Green Chemistry, 16 (2014) p. 1865–1870

[25] N. von der Assen, P. Voll, M. Peters, A. Bardow, Chem. Soc. Rev. 2014, DOI: 10.1039/c3cs60373cN

[26] Farbe u. Lack 120 (2014) 10, p. 7, Market Study of Epoxy Resin by Application and Geography – Trend and Forecast to 2019 Markets and Markets

[27] R. Lässig, M. Eisenhut, A. Mathias, R.T. Schulte, F. Peters, T. Kühmann, T. Waldmann, W. Begemann. Serienproduktion von hochfesten Faserverbundbauteilen – Perspektiven für den deutschen Maschinen- und Anlagenbau, Roland Berger Strategy Consultants, 09/2012

[28] E. Witten (AVK), T. Kraus, M. Kühnel (CCeV), Composites – Market Study 2014, Market Trends, Outlook and Chalenges. Carbon Composites and AVK, 6th of Oktober 2014

Authors

Prof. Dr. Michael Dornbusch studied chemistry at the Heinrich-Heine University in Düsseldorf/Germany, obtaining his doctorate in organic chemistry in 2001. He then did a two-year post-doc at the Max-Planck Institute for Iron Research, where he clarified the mechanism of film formation in modern conversion layers. In 2003, he took charge of the Corrosion Lab at BASF Coatings AG, Münster/Germany. His job there entailed the development of new corrosion-protection systems and establishing electrochemical methods for the coatings industry. In 2009, he changed to Dörken MKS GmbH, where he headed the R&D department for zinc-containing corrosion-protection coatings and friction coatings. Since 2011, he has been lecturing at the University for Applied Sciences, Krefeld/Germany, in the fields of coatings technology, electrochemistry and corrosion. His research focus is on corrosion protection and corrosion processes.

Dr. Ulrich Christ studied Chemistry at the University of Stuttgart/Germany with a focus on polymer chemistry. He obtained his PhD at the Forschungsinstitut für Pigmente und Lacke (FPL, Research Institute for Pigments & Coatings) and at the University of Stuttgart. He worked over 20 years in the R&D in the chemical industry, first as head of the research and applied technology of pigments and silicas for coatings at Evonik (formerly Degussa) and thereafter as head of R&D of industrial coatings at FreiLacke. In 2008, he came back to the FPL, which was integrated to the Fraunhofer Institute IPA in Stuttgart in 2009 under founding the department coating systems and painting technology. There he is responsible for the group coatings chemistry and applied technology. His main fields of activity are the corrosion protection and short-term tests of coatings as well as studying structure-property relationships of pigments, polymeric binders and additives in organic coatings.

M. Dornbusch, U. Christ, R. Rasing: Epoxy Resins

Robert M. T. Rasing graduated at the Radboud University of Nijmegen/The Netherlands, in 1999 with a major in physical organic chemistry. After formulating with two-component polyurethane and amine-cured epoxides at NKC Deventer, The Netherlands, he joined Air Products, Inc. in 2000. At Air Products, Rob specialized in the applied technology of amine curing agents for epoxy resins and polyisocyanates and their use in products for the construction ("civil engineering") and coatings industries. He is co-inventor on patents in the field of application and a frequent speaker at international conferences and seminars. Currently, Rob holds the position of European Technology Manager Epoxy and is based in Utrecht/The Netherlands.

Index

Symbols

A

B

C

D

E

F

G

H

I

K

L

M

N

O

P

Q

R

S

T

U

V

W

Z

www.ingramcontent.com/pod-product-compliance
Lightning Source LLC
LaVergne TN
LVHW021727021026
849161LV00004BA/762